T0297347

Applications of Graphene and Graphene-Oxide Based Nanomaterials

Applications of Graphene and Graphene-Oxide Based Nanomaterials

Sekhar Chandra Ray
Department of Physics, College of Science,
Engineering and Technology, University of South Africa,
Private Bag X6, Florida, 1710, Science Campus, Christiaan de Wet and
Pioneer Avenue, Florida Park, Johannesburg, South Africa.

ELSEVIER

AMSTERDAM • BOSTON • HEIDELBERG • LONDON
NEW YORK • OXFORD • PARIS • SAN DIEGO
SAN FRANCISCO • SINGAPORE • SYDNEY • TOKYO
William Andrew is an imprint of Elsevier

William Andrew is an imprint of Elsevier
225 Wyman Street, Waltham, MA 02451, USA
The Boulevard, Langford Lane, Kidlington, Oxford, OX5 1GB, UK

Copyright © 2015 Elsevier Inc. All rights reserved.

No part of this publication may be reproduced or transmitted in any form or by any means, electronic or mechanical, including photocopying, recording, or any information storage and retrieval system, without permission in writing from the publisher. Details on how to seek permission, further information about the Publisher's permissions policies and our arrangements with organizations such as the Copyright Clearance Center and the Copyright Licensing Agency, can be found at our website: www.elsevier.com/permissions

This book and the individual contributions contained in it are protected under copyright by the Publisher (other than as may be noted herein).

Notices
Knowledge and best practice in this field are constantly changing. As new research and experience broaden our understanding, changes in research methods or professional practices, may become necessary.

Practitioners and researchers must always rely on their own experience and knowledge in evaluating and using any information or methods described herein. In using such information or methods they should be mindful of their own safety and the safety of others, including parties for whom they have a professional responsibility.

To the fullest extent of the law, neither the Publisher nor the authors, contributors, or editors, assume any liability for any injury and/or damage to persons or property as a matter of products liability, negligence or otherwise, or from any use or operation of any methods, products, instructions, or ideas contained in the material herein.

ISBN: 978-0-323-37521-4

Library of Congress Cataloging-in-Publication Data
A catalog record for this book is available from the Library of Congress

British Library Cataloguing-in-Publication Data
A catalogue record for this book is available from the British Library

For Information on all William Andrew publications
visit our website at http://store.elsevier.com/

This book has been manufactured using Print On Demand technology. Each copy is produced to order and is limited to black ink. The online version of this book will show color figures where appropriate.

CONTENTS

ACKNOWLEDGMENTS

I would like to thank Professor Nikhil Ranjan Jana of Centre for Advanced Materials, Indian Association for the Cultivation of Science (IACS), Jadavpur, Kolkata, India, for his contribution to the third chapter entitled *"Graphene-Based Carbon Nanoparticles for Bioimaging Applications"*. I also would like to thank my wife Ms Susmita Ray and son Master Shrishmoy Ray for their encouragement for writing this book.

ACKNOWLEDGMENTS

CHAPTER 1

Application and Uses of Graphene

Sekhar C. Ray

Department of Physics, College of Science, Engineering and Technology, University of South Africa, Florida Park, Johannesburg, South Africa

1.1 INTRODUCTION

Carbon has many different forms, namely diamond, graphite, and amorphous carbon. Diamond and graphite are well-known allotropes of carbon known since ancient times. Fullerene, the third form of carbon, was discovered in 1985 by Kroto et al., and carbon nanotubes (CNTs) were discovered in 1991 by Iijima; subsequently, it has become very important in the science and technology communities. Thus, only three-dimensional (3D) (diamond and graphite), one-dimensional (1D; CNTs), and zero-dimensional (0D; fullerenes) allotropes of carbon were known in the carbon community. Although it was realized in 1991 that CNTs were formed by rolling of a two-dimensional (2D) graphene sheet, with a single layer from 3D graphitic crystal, the isolation of graphene was quite elusive, resisting any attempt regarding its experimental research work until 2004. Graphene is the basic structural element of some carbon allotropes, including graphite, CNTs, and fullerenes. Fullerene is entirely composed of carbon in the form of spherical shapes called bucky balls, whereas CNTs have tubular arrangements. For more than two decades, fullerene and CNTs-based materials enjoyed widespread applications in diverse fields of research such as electronics, batteries, super-capacitors, fuel cells, electrochemical sensors, biosensors, and medicinal applications.

Currently, graphene is becoming a "rising star" material after its successful production by a simple scotch tape approach using readily available graphite in 2004 by Andre Geim and his coworkers. Graphene comprises a single-layer sheet of sp^2 bonded carbon atoms with densely packed honeycomb crystal lattice. Its exceptional properties such as high surface area, room temperature Hall effect, tunable band gap, and excellent electrical, thermal, and conducting properties offer a versatile platform for its use as active material in the preparation of various composite materials (Novoselov et al., 2004). Numerous efforts were made to review the structure, preparation, properties, and applications of graphene and its composite materials (Geim, 2009; Rao et al., 2009; Neto et al., 2009; Allen et al., 2010). Currently, graphene is one of the most popular materials; it can be applied for various devices and applications due to its outstanding properties. This chapter presents the different uses and applications of graphene, and the synthesis process and different outstanding properties are also discussed briefly.

1.2 PREPARATION/SYNTHESIS OF GRAPHENE

Several methods have already been established for producing different kinds of graphene materials. Micromechanical exfoliation, chemical vapor deposition, epitaxial growth, arc discharge method, intercalation methods in graphite, unzipping of CNTs, and electrochemical and chemical methods were some of the important preparation methods available for the preparation of graphene. Chemical methods involve strong oxidation of graphite and subsequent reduction to graphene by reducing agents. A novel synthesis by dichromate oxidation of graphite followed chemical reduction with hydrazine, which is also used for the preparation of graphene (Chandra et al., 2010). Kumar et al. (2013) reported the preparation of nitrogen-doped graphene by microwave plasma chemical vapor deposition method. Electrophoretic deposition (EPD) is one of the interesting techniques for synthesizing a nanosheet of graphene. Chen et al. (2010) deposited graphene sheets onto nickel foams via EPD approach. Ata et al. (2012) prepared graphene by EPD method with aluminon as an organic charging and film-forming agent. Graphene could be prepared by direct current arc-discharge method in the presence of hydrogen atmospheric pressure using graphite rods as electrodes for the deposition (Guo et al., 2012). Laser pyrolysis technique has been demonstrated to synthesize multilayer graphene in the presence of dilution gas (Florescu et al., 2013). Among these methods, chemical vapor deposition methods (plasma-enhanced CVD/thermal CVD) are efficient approaches for the synthesis of graphene. However, each method has its own advantages and disadvantages. Among all of these methods, CVD method is efficient for the production of graphene materials for different applications. Graphene, produced in this method, was found to have better crystallinity than that formed with any other method. PECVD method has shown the versatility of synthesizing graphene on any substrate, thus expanding its field of applications.

1.3 PROPERTIES OF GRAPHENE

The bond length of the C−C bond in graphene is ∼1.42 Å, with a strong bond in a particular layer but weak bonding between layers. The specific surface area of a single sheet of graphene is ∼2630 m^2/g (Stoller et al., 2008). Graphene has unique and outstanding optical properties (>97.7% transmittance), with a band gap value of ∼0−0.25 eV (Zhang et al., 2009). Some other fascinating characteristics include high carrier mobility (∼200,000 cm^2/Vs) (Geim and Novoselov, 2007) and high Young's modulus (1.0 TPa). Graphene and its composite materials can be used as semi-conductors because of their extraordinary conducting properties. Graphene has been envisioned as the building block of all other important graphitic allotrope forms: fullerene-wrapped version of graphene; CNT-rolled version of graphene; and graphite-stacked version of graphene. Enoki et al. (2005) investigated the unique magnetic properties of nanographene, such as spin glass states, magnetic switching, and edge-state spin gas probing, for the possible applications in electronic and magnetic devices. Chen, S. et al. (2012) reported brief experimental studies about the isotope effects on the thermal properties of graphene and found that the ratios of ^{12}C and ^{13}C play an important role in the thermal conductivity of graphene (Chen, S. et al., 2012). All these extraordinary properties of graphene have led to its inherent use for real applications.

Some of the potential properties are as follows:

- High Young's modulus ∼1000 Gpa
- Effective moisture barrier
- Electrical conductivity similar to copper
- Density four-times lower than copper
- Thermal conductivity five-times that of copper
- Essentially an opened up CNT; high surface area of ∼2500 m^2/g
- Lower density than steel but can be up to 50-times stronger

However, for a quick reference regarding the synthesis of graphene during different processes and grown/synthesized on different substrates, studies of their different properties using different measurements

Table 1.1 Advantages and Disadvantages for Techniques Currently Used to Produce Graphene

	Advantages	Disadvantages
Mechanical exfoliation	(i) Low cost and easy (ii) No special equipment needed, (iii) SiO$_2$ thickness is tuned for better contrast	(i) Serendipitous (ii) Uneven films (iii) Labor intensive (not suitable for large-scale production)
Epitaxial growth	(i) Most even films (of any method) (ii) Large-scale area	(i) Difficult control of morphology and adsorption energy (ii) High-temperature process
Graphene oxide	(i) Straightforward upscaling (ii) Versatile handling of the suspension (iii) Rapid process	(i) Fragile stability of the colloidal dispersion (ii) Reduction to graphene is only partial

Source: *Reproduced with permission (Soldano et al., 2010). Copyright 2015 Elsevier.*

including their advantages/disadvantages are listed in Tables 1.1 and 1.2 (Soldano et al., 2010).

1.4 POTENTIAL APPLICATION AND USES OF GRAPHENE

Graphene has already demonstrated high potential to impact most information communication technology areas, ranging from top-end high-performance applications in ultrafast (>1 THz) information processing to consumer applications using transparent or flexible electronic structures. The great promise of graphene is evidenced by the increasing number of chip-makers now active in graphene research. Most importantly, graphene is considered to be among the candidate materials for post-Si electronics. Potential applications of graphene include electronics, light processing, energy storage and generators, sensors, plasmonics, and meta-materials, as well as various medical and other industrial processes enhanced or enabled by the use of new graphene materials. An overview of graphene applications is presented in Figure 1.1 (Brownson et al., 2011; Hong et al., 2011; Ramachandran et al,. 2013; Singh et al., 2011).

1.4.1 Graphene in Hydrogen Storage Devices

Hydrogen is the main contender and an abundant element for developing next-generation clean fuel. Hydrogen-based fuel cells are promising solutions for the efficient and clean delivery of electricity. Hydrogen can react with ambient oxygen to release energy, leaving only water as a waste product. Overcoming the main hurdles regarding

Substrate		Growth Condition (gas, T, exposure)	Experiment Techniques	Edges	Comment ($a_C = 0.245$ nm)	Reference
Metals	Pt(111)	Benzene (C_6H_6), $T = 1000$ K. 1–5L → nongraphitic film; >5L → full coverage	STM, LEED, AES	Hexagonal arrangement beyond edges		Fujita et al. (2005)
		Ethylene (C_2H_4), $T = 800$K. 5L exposure (If $T > 1000$K → graphitic island)	LEED, STM	No clear hexagonal arrangement; No growth over the edges	$a_{P\text{-}t} = 0.278$ nm $a_{Moire} \rightarrow 2.2$ nm	Land et al. (1992)
		HOPG on 1ML graphitic film	AFM, PCM	Continuous film from upper terrace to lower terrace	0.738 nm $< a < 2.1$ nm	Enachescu et al., 1999
	Pt(755)	Chemical vapor deposition	LEED, XPS, ARUPS	Formation of large sheet		Ryoko et al. (2000)
	Ni(111)	Chemical vapor deposition	LEED, AES vibrational spectro		Evidence of Fuchs–Kliewer phonons	Tanaka et al. (2003)
	Ni(110)	Carbon monoxide (CO), $T = 600$ K. 90,000L exposure	SEELFS		Graphitic layer on (110) faces	Papagno et al. (1984)
	Ru(001)	Ethylene (C_2H_4). T-dependent solubility gradient	LEEM, SEM, μ-Raman, AES, electrical	No growth "uphill" over the edges	$a_{Ru} = 0.271$ nm $a = 0.145$ nm (1st layer) $a_{Moire} = 3$ nm	Sutter et al. (2008)
	Ir(111)	Ethylene (C_2H_4), $T > 1100$ K. 1L exposure	STM	Growth beyond both sides of the edges	$a_{Ir} = 0.272$ nm $a_{Moire} = 2.5$ nm	Coraux et al. (2008)
	Co(0001)	Acetylene (C_2H_2), $T = 410$K. 0.6L–3.6L exposure	XPS, XPD, LEED, TDS, LEIS		K enhances the coverage of the surface	Vaari et al. (1997)
Carbides	nH–SiC ($n = 1,2,...$)	Si sublimation, $T \sim 1670$ K.	LEED, X-ray, STM	Formation of large continuous sheet over terraces		Charrier et al. (2002)
	TiC(111)	Chemical Vapor Deposition on faceted surface, $T = 1770$ K	XPS, ARUPS, LEED	No edge-localized state	Growth on each facet	Terai, M., et al. (1998)
	TiC(410)	Chemical Vapor Deposition on platelets surface, $T = 1770$ K	XPS, ARUPS, LEED	No growth over the edge	Nanoribbon growth (≈ 1–2 nm)	Terai, M., et al. (1998)
	TaC(111)	Ethylene (C_2H_4), $T = 1570$ K 100,00L exposure, $T = 1270$ K	AES, LEED, STM	Coverage is interrupted at terrace interface	$a = 0.249$ nm (1st layer) $a = 0.247$ nm (2nd layer)	Nagashima et al. (1994)

"Acronyms are as follows: L, single-layer coverage; ML, monolayer; T, temperature; STM, scanning tunneling microscopy; LEED, low-energy electron diffraction; AES, Auger electron spectroscopy; AFM, atomic force microscopy; PCM, point-contact microscopy; XPS, X-ray photoemission spectroscopy; ARUPS, angle-resolved ultraviolet photoelectron spectroscopy; SEELFS, surface-extended-energy-loss fine structure; SEM, scanning electron microscopy; XPD, X-ray photoelectron diffraction; TDS, time-domain spectroscopy; LEIS, low-energy ion spectroscopy.
Source: Reproduced with permission (Soldano et al., 2010). Copyright 2015 Elsevier.

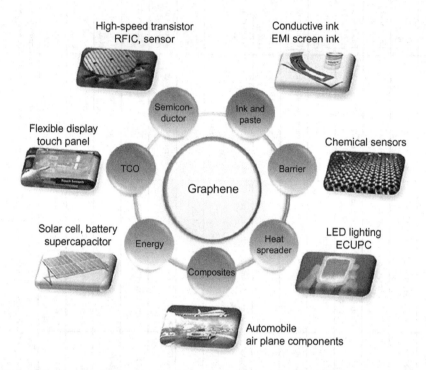

Figure 1.1 Overview of applications of graphene. Reproduced with permission (Hong, 2011).

practical applications and the problem of hydrogen storage and transport, particularly for use in mobile applications, must be accomplished before hydrogen can become a widespread energy source. Because hydrogen is an energy carrier, a key step for the development of a reliable hydrogen-based technology requires solving the issue of storage and transport of hydrogen. In this respect, the novel material graphene (a single sheet or a few sheets of graphite) has recently attracted attention as a promising storage medium. Graphene has an extremely high specific surface area and, in combination with its low weight, robustness, chemical inertness, and attractive physicochemical properties, is among the most suitable materials for hydrogen storage. Hydrogen can react with graphene surfaces by the process of physisorption (Van der Waals forces) and chemisorption (forming a chemical bond with C atoms), and the efficiency of storage is usually measured by two parameters: the gravimetric density (GD), namely the weight percentage of hydrogen stored in the total weight of the system, and the volumetric density (VD), which is the stored hydrogen mass per unit volume of the system. The interaction between hydrogen and graphene can be tuned by adjusting the distance between adjacent layers or by chemical functionalization of the material that could be tuned for the enhancement of adsorption/desorption properties of hydrogen on graphene. The binding of molecular hydrogen is weak and therefore requires low temperatures and high pressure to ensure reasonable storage stability. In the most favorable conditions (high pressure and low temperature), H_2 can form a uniform compact monolayer on the graphene sheet. The chemisorption of atomic hydrogen is a favorable process because H binding energy is higher than chemisorption barriers to energy. The formation of "dimers" of H on the graphene surface brings higher energy with respect to isolated bound H (Ferro et al., 2008). Atomic hydrogen absorption on epitaxial graphene on SiC also allows formation of dimer structures, preferential adsorption of protruding graphene areas, and clustering at large hydrogen coverage (Balog et al., 2009; Guisinger et al., 2009). The maximum GD reachable in graphene with chemisorption is 8.3% (1/12), which corresponds to the formation of a completely saturated graphene sheet with 1:1 C vs. H stoichiometry, called *graphane* (Sofo et al., 2007). In case of physisorption, the VD depends on the possibility of building compact structures with *graphene* (or *graphane*) sheets. The energy profiles for the processes of adsorption of hydrogen on graphene are summarized in Figure 1.2. Graphene is a single-layer quasi-2D system and its VD is not well-defined, so 3D multi-layer graphene in the evaluation of the potentialities of hydrogen storage devices should be considered. The pillaring method is a process by which a layered intercalated structure is converted into thermally stable, meso-porous material with a large surface area. A theoretical approach by the group of Froudakis (Dimitrakakis et al., 2008) proposes a novel 3D material (Bendikov et al., 2004) consisting of graphene layers connected by CNTs acting as "pillars" that stabilize the structure and keep the graphene layers at a fixed distance (Dimitrakakis et al., 2008).

It was observed by Patchkovskii et al. (2005) from theoretical calculations that the physisorption energy is nearly doubled with respect to the monolayer graphene, reaching values of ~ 0.1 eV, because the attractive VdW forces of two layers combine together. In this case, the GD is increased by ~ 30–40% of its single-layer value, potentially reaching 8% at high pressure and low temperature but remaining in the range of 3–4% at room temperature and high pressure. It was also confirmed that the GD value of 3% could be reached at room temperature only with high pressure, whereas at low temperature much higher values can be obtained. From the experimental side, it was shown that

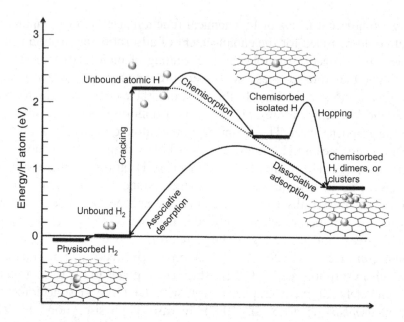

Figure 1.2 Energy level diagram for the graphene−hydrogen system. The energy is in eV per H atom (i.e., to obtain the values per H_2, each energy and barrier value must be doubled). Values of energy levels and barriers are deducted from the experimental and theoretical evaluations, with average values taken when different values are available. The reference level is the pristine graphene plus unbound molecular hydrogen. Reproduced with permission (Tozzini et al., 2013). Copyright 2015, RSC publication.

such a layered structure can be realized by using graphene oxide and the interaction between hydroxyl groups and boronic acids (Burress et al., 2010).

In the hydrogen storage process in graphene, the molecular hydrogen adsorption (or GD and GV) at room temperature/ambient pressure conditions could be improved through chemical functionalization or suitably modifying the surface through doping or adsorption of metals (alkali metals, alkaline earth metals, transition metals), as reported by many theoretical investigations. In light of the weak interaction between molecular hydrogen and graphite/graphene, several routes were investigated to enhance both the binding and the gravimetric/volumetric storage capacity. One approach exploits the chemical decoration of graphene with alkali atoms such as Li, Na, and K. In the case of Li, it was shown that each adsorbed Li on graphene and nanostructured graphene can adsorb up to four H_2 molecules, amounting to GD more than 10 wt % (Ataca et al., 2008; Du et al., 2010). A change in hydrogen binding energy on K-decorated graphene was also predicted (Tapia et al., 2011).

The advantages for hydrogen storage of this particular chemical inter-action, half-way between chemisorption and physisorption, are related to optimization of hydrogen–graphene interaction for room tempera-ture application. The chemical functionalization of graphene with different transition metals such as Sc, V, and Ti (Durgun et al., 2008; Liu, C. et al., 2010; Liu, Y. et al., 2010; Liu, Z. et al., 2010; Kim et al., 2008) also improve the storage capacity of hydrogen in graphene. The Pd-decorated graphene increases the GD at room temperature (30 bar pressure) from 0.6% to 2.5% (Parambhath et al., 2011). Theoretical studies have proposed that the Ca-decorated graphene nanoribbons (GNRs) for hydrogen adsorption reaching gravimetric capacities of 5% with negligible clustering of Ca atoms (Lee et al., 2010). A few approaches to hydrogen storage in graphene are discussed mostly on theoretical grounds. For those that possess sufficiently large storage capacity, heat must be applied to release hydrogen. In the case of applications to vehicles, this would involve the use of an onboard burner and heat exchanger. Concerning hydrogen storage, one of the key problems to be solved for practical applications is related to the realization of chemisorption/desorption mechanism that works at room conditions, without the recourse to extreme temperatures and pressures. The chemisorb process is relatively easy, at least for atomic hydrogen, but the desorption process requires overcoming the associative barrier that is of the order of eV. To overcome this problem, Tozzini et al. (2013) recently proposed exploiting the control of the curvature of graphene to desorb hydrogen. In this process, H atoms are loaded in the graphene structure after lateral compression of graphene that preferentially binds on convexities, as was explained by Tozzini et al. (2013). In this configuration, C–H binding is very stable on graphene and can be safely transported with minimal or no dispersion.

1.4.2 Graphene as a Battery

Graphene proves to be an extremely interesting and innovative material in portable energy storage devices. Batteries are extensively used in automobiles (cars and motor bikes), aircraft, boats, ships, and elec-tronic equipment (Zhang et al., 2008; Chen, D. et al., 2012; Chen, S. et al., 2012; Chen, X. et al., 2012; Chen, Y. et al., 2012). Conventional lithium–ion batteries are one of the promising energy storage devices that can be used in portable electronic applications (Chen et al., 2011). The current focus on rechargeable batteries has made reasonable improvements in terms of higher capacity and compact size for convenient usage. Graphene and its composite materials were used as novel electrode materials for lithium−ion battery applications (Atabaki et al., 2013). The specific energy of Li−C batteries is quite low (370 mAh/g) (Liang et al., 2009) because six carbon atoms can host only one lithium ion by forming an intercalation compound (LiC_6).

The excellent properties of graphene and ease of fabrication toward preparation of graphene-based composites with metal, metal oxides, and polymers make them extraordinary materials in the field of batteries. Li−ion batteries comprising $LiFePO_4$ as the cathode and TiO_2/graphene composite as the anode have been demonstrated by Choi et al. (2010) with negligible degradation, even after 700 cycles at 1 C_m. A lilly-like graphene sheet-wrapped nano-Si composite was prepared by spray-drying processes that afforded the reversible capacity value of 1525 mAh/g (Hey et al., 2011). Mai et al. (2011) prepared CuO/graphene composite by *in situ* chemical method, which exhibited the reversible capacity of 583.5 mAh/g, and the reversible capacity retained capacity of 75.5% even after 50 cycles, proving the excellent performance of the prepared composite for applications in Li−ion batteries (Mai et al., 2011). $CoFe_2O_4$−graphene nanocomposite has been prepared by hydrothermal method and used as anode material for lithium−ion batteries (Xiao et al., 2012). In Li−ion batteries, silicon, tin, and metal oxides are used as electrodes, but these are extremely specific high-energy electrodes that suffer from large-volume expansion−contraction during the charge−discharge process known as irreversible cracking and crumbling (pulverization) of the anodes (Ju et al., 2010). To overcome this problem, several researchers are trying to use graphene to create silicon or metal oxide composites that show higher specific energy and less variations in their physical properties. Wang (2010) produced freestanding graphene−Si nanocomposite films by an in situ chemical method. Silicon can reversibly accommodate lithium by forming $Li_{4.4}Si$ alloys, and graphene can minimize volume expansion that leads to pulverization. This new electrode shows a capacity of ∼708 mAh/g, even after 100 cycles. The durability of these films is due to the graphene−Si void spaces that buffer the volume change that would normally occur in Si electrodes (Wang 2010). The CuO/graphene nanocomposite exhibited a reversible capacity of 600 mAh/g, even after 100 cycles. A promising anode is represented by <20 nm anatase TiO_2 nanoparticles coated onto graphene sheets. Used in combination with a $LiFePO_4$ cathode, the battery can deliver

100% coulombic efficiency even after 700 cycles at 1 C_m (measured C rate), except during the initial few cycles when irreversible loss was observed (Choi et al., 2010).

1.4.3 Application of Graphene Thin Film as Transparent Conductor (Electrodes)

Graphene is an elastic/flexible thin film that behaves like transparent plastic; it conducts heat and electricity better than any metal. It behaves as an impermeable membrane and is chemically inert and stable. Thus, graphene seems the "ideal" for the production of next-generation transparent conducting electrodes required for different applications such as solar cells, light-emitting diodes (LEDs), organic light-emitting diodes (OLEDs), touch screens, smart windows and liquid crystal displays (LCD), and organic photovoltaic cells (OVPs) (Bonaccorso et al., 2010). There is a real need to find a substitute for indium tin oxide (ITO) in the manufacturing of various types of displays and touch screens due to the scarcity of indium and its consequent growing cost. In particular, graphene's mechanical strength and flexibility are advantageous compared with ITO, which is brittle. Thus, coupled with carbon's abundance, graphene presents a more sustainable alternative to ITO for these applications.

Figure 1.3 shows possible applications of flexible transparent electrodes for corresponding sheet resistance ranges. Most of the industrial

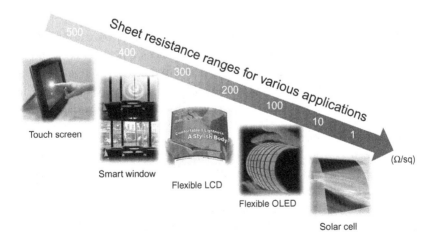

Figure 1.3 Potential graphene applications and corresponding resistance range required for each application. Reproduced with permission (Bae, S. et al., 2012). Copyright 2015 IOP publication.

applications require sheet resistance less than 500 Ω/sq for transparent electrodes. For example, the sheet resistance for OLED displays and solar cells should be less than ~50 Ω/sq, and for touch screens a sheet resistance of 200–500 Ω/sq is acceptable within transparency >90%.

1.4.3.1 Graphene as Transparent Conducting Electrodes

Thin-film conducting electrodes have attracted considerable attention because of their vast applications in the field of thin-film transistors that are used in plastic electronics such as large-area flat panel displays, radio frequency identification tags, and smart cards. The 2D single atomic layer graphene possesses work functions (4.7–4.9 eV) similar to that of Au, forms low contact resistance with organic materials, has high crystallographic quality, and has ballistic electron transport on the micrometer scale, with only 2.3% of light absorption that could be used in transparent conducting electrodes (Basu et al., 2014), in fabricated graphene-based electrodes, and in organic thin-film transistors (OTFTs). Basu et al. (2014) found that the pentacene OTFTs with graphene source–drain (S–D) electrodes exhibits superior performance with a mobility of 0.1 cm^2/V/s and on–off ratio of 10^5 compared with OTFTs with Au-based S–D electrodes. They also studied the feasibility of using graphene as an S–D electrode in OTFTs with minimal contact resistance, which provides an alternative material with high carrier efficiency, chemical stability, and excellent interface properties with organic semi-conductors. This finding confirmed the potential of replacing expensive Au electrodes.

1.4.3.2 Flexible Electronics

Graphene and graphene-related 2D crystals and hybrids will have a disruptive impact on current optoelectronics devices based on conventional materials, not only because of cost/performance advantages but also because they can be manufactured in more flexible ways suitable for a growing range of applications such as touch screens and flexible displays.

1.4.3.3 Touch Screen

Touch screen is another potential application of graphene transparent conductive films that have been adopted in various electronic devices such as cell phones, laptops, smart phones, e-books, and others. Researchers in Korea and Japan have fabricated films of graphene—planar sheets of carbon that are one atom thick—measuring tens of centimeters. The researchers engineered these large graphene films into transparent electrodes, which were incorporated into touch screen panel devices. Different types of commercialized touch screens are available in the market. Resistive-type touch screens are operated through induction of an electric short between top and bottom transparent conducting films. They require a resistance of up to 550 $\Omega/\square$ and an optical transmittance of more than 90% at a 550-nm wavelength. Typically, ITO films have been widely used in these applications. Bae et al. (2010) reported layer-by-layer graphene stacking to fabricate a doped four-layer graphene thin film on flexible PET substrates and found its sheet resistance as low as ~30 $\Omega/\square$ with ~90% transparency at a 550-nm wavelength, which is superior to commercial transparent electrodes such as ITO that could be used in a fully functional touch screen panel device capable of withstanding high strain. In this fabrication, all of the materials are flexible and the screen operates reliably after many bending cycles. In addition, capacitive touch screens fabricated using transparent graphene films are also available because graphene films can satisfy the requirement of a sheet resistance value of <100 $\Omega/\square$ through doping (Lee et al., 2013). They have predicted that in the future, this transparent graphene films would replace rigid and brittle ITO films in touch screen electrodes.

1.4.4 Solar Cells and OVPs

Solar cells (polymer, bulk hetero-junction, and dye-sensitized) are the most important promising devices for the conversion of sunlight into electrical energy (Bundgaard et al., 2010); they offer the advantages of low cost and large-scale production. As the economies of scale reduce the production costs, rapid growth of the PV industry causes the depletion of the raw materials involved in the production of solar panels. Most of the solar cells are ITO with a nonconductive glass protective layer to meet their needs. OPV devices are made of an organic layer sandwiched between two charge-collecting electrodes, one of which must be transparent (e.g., ITO or fluorine tin oxide [FTO]) and the other is usually aluminum, sometimes coated with LiF or MgO. Efforts were made to replace these materials by finding alternative electrode materials for the fabrication of solar cell applications. Indium is extremely rare, so it is becoming more expensive. This will keep solar cells expensive in the future, whereas graphene could be very cheap because carbon is abundant. However, ITO is a nonflexible material and graphene and graphene-based nanocomposite are not only flexible but also extremely conductive and could be very useful in solar cell devices. Wu, M.S. et al. (2013), Wu, S. et al. (2013), and Wu, Y. et al.

(2013) achieved photoelectron conversion efficiency of ~7.5% using graphene nanosheets (GNSs) incorporated with activated carbon as the counter electrode. Molybdenum sulfide (MoS_2)/graphene composite electrodes have been used as a counter electrode for dye-sensitized solar cells (DSSCs) (Yue et al., 2012). The composite-modified electrode has higher current density than those of MoS_2. This composite exhibited a power efficiency value of 5.98%, which was comparable with that of the Pt electrode (6.23%), and revealed good performance of the molybdenum-based composite electrode. The fabrication of low-cost DSSCs achieves high efficiency.

1.4.4.1 Organic Photovoltaic Cells

OPVs have few advantages over classical solar harvesting devices: (i) they are flexible and semi-transparent; (ii) they are manufactured in a continuous printing process by coating large areas; (iii) they are integrated in different devices; and (iv) they are cost-effective and environmentally friendly. The same challenges that are efficiency, lifetime, and competitive substitutes for ITO is graphene. Graphene was found to be an alternate to ITO in OPVs because of its unique sheet resistance, high charge collection efficiency, as well as transparency. To achieve this goal, graphene films with low electrical resistance and high work function must be developed. The sheet resistance and work function of graphene can be tuned by the number of layers and doping to produce highly efficient OPV. Many researchers have worked on applying graphene films as counter electrodes for transparent electrodes for OPVs (Wang et al., 2008, 2011; Yin et al., 2010; Arco et al., 2010). Zhang (2010) fabricated OVP cells composed of CVD graphene films and found that the power conversion efficiency (PCE) was very high; this can be attributed to a higher optical transmittance and conductance. Park et al. (2010) also grew graphene sheets by CVD to replace ITO as the negative electrode, with PCE comparable with devices containing ITO (1.63% for doped graphene vs. 1.77% for the latter). As indicated in other work by Liu, C. et al. (2010), Liu, Y. et al. (2010), and Liu, Z. et al. (2010), graphene could be used not only to replace ITO but also to enhance electron transport and exciton dissociation in the heterojunction of a solar cell. They questioned the use of [6,6]-phenyl-C_{61}-butyric acid methyl ester (PCBM) as the standard electron acceptor, and combined solution-processable functionalized graphene (SPF graphene) and functionalized multiwalled carbon nanotubes (f-MWCNTs) to produce a new active layer that was then sandwiched between

PEDOT:PSS and LiF. The best result obtained was a PCE of 1.05% (Liu et al., 2008). Other types of promising, cost-effective, flexible solar cells alternative to OPVs are DSSCs, in which light-absorbing dye provides electrons that are collected from the TiO_2 support layer and travel to the external circuit. DSSCs have recently reached efficiencies that are comparable with those of amorphous Si cells. Gratzel, 2007 and Wang et al., 2008 tried to replace ITO and FTO with graphene as window electrodes to simplify fabrication and lower the cost of production, even though PCE is still much lower as compared with standard devices.

1.4.5 Fuel Cells

Fuel cells are a type of energy storage device that converts chemical energy from a fuel into electrical energy by using oxygen and methanol. The green energy of fuel cells could be obtained from reduction of oxygen (Qu et al., 2004) and oxidation of methanol (Golabi et al., 2002). Most of the electrocatalytic performance is based on their selection in suitable electrode materials. Numerous efforts were made in the literature for the utilization of carbon-based composite materials toward oxygen reduction reactions (Wu, J. et al., 2011; Wu, S. et al., 2011). One of the main issues connected with fuel cells is the limited availability of platinum (Pt), which is a candidate catalyst for fuel cell reactions. After the discovery of graphene, it was found to have widespread applications in fuel cells, where it can be used as an excellent electrode material because of its excellent physicochemical properties. Graphene has a high surface area, making it more efficient than carbon black for dispersing Pt nanoparticles (Xin et al., 2011). However, Pt acts as an active cathode material and can also be deposited onto other electrode materials to improve their electrocatalytic properties toward methanol oxidation such as Pt oxides, Pt-Sn, and Pt-Ru (Hu et al., 1999; Sobkowski et al., 1985; Bell et al., 1998). The new Pt/graphene catalyst used in direct methanol fuel cells (DMFCs) shows: (i) enhanced interactions between Pt and graphene; (ii) additional Pt active sites; (iii) less defects on graphene, thus improving the stability of graphene; and (iv) better ordered Pt surface morphology, thus introducing more active catalytic sites. Kou et al. (2009) have synthesized Pt nanoparticle-supported functionalized graphene sheets for the electrocatalytic reduction of oxygen. The composite has good catalytic activity and better stability in both electrochemical surface area and oxygen reduction activity. Zhang et al. (2012) have synthesized graphene/polyallylamine−Au nanocomposites and exploited them for the electrocatalytic reduction of oxygen. GNSs also have been considered for polymer electrolyte fuel cells (PEFCs) because of their higher carbon monoxide (CO) tolerance (Yoo et al., 2011). The GNS-CNT hybrid nanostructure provides numerous edge planes with strong electrochemical activity for the achievement of good performance in fuel cell applications (Du et al., 2012). Graphene-supported Pt electrocatalyst was prepared for methanol oxidation (Kakaei et al., 2013).

1.4.6 Microbial Biofuel Cells

Microbial fuel cells are a type of fuel cell utilizing microorganisms to produce electricity from organic wastes. Graphene/carbon cloth, graphene/PANI, graphene-modified stainless steel mesh, and crumpled graphene electrodes have been demonstrated as microbial fuel cells (Liu, J. et al., 2012; Liu, M. et al., 2012; Hou et al., 2013; Zhang et al., 2011; Xiao et al., 2012).

1.4.7 Enzymatic Biofuel Cells

Recently, there has been substantial interest toward the development of enzymatic biofuel cells (EBFCs) because they possess the potential to be used as an in vivo power source for implantable medical devices such as pacemakers (Liu, C. et al., 2010; Liu, Y. et al., 2010; Liu, Z. et al., 2010). Due to graphene's excellent conductivity, ballistic electron mobilities at room temperature (Liu, C. et al., 2010; Liu, Y. et al., 2010; Liu, Z. et al., 2010), large surface area, and other unique properties (as discussed), it is thought of as an optimal replacement and starting platform for further research; high-performance EBFCs are expected soon. Although this is a relatively new area of graphene research and, consequentially, there is limited literature available, one of the few articles currently available is highly exciting. The most compelling advancement within this area of electrochemistry concerns the use of GNSs within the construction of membraneless EBFCs, as reported by Liu, J. et al. (2012) and Liu, M. et al. (2012).

1.4.8 Organic Light-Emitting Diodes

An important component of OLEDs is the anode as transparent electrode material, injecting charge carriers and allowing light to pass through. Graphene films are attractive materials for flexible transparent conductive electrodes in OLEDs due to their controllable transparency, good electrical conductivity, and suitably tunable work function. In particular, graphene films have a molecular structure similar to that of organic electronic materials, and thus can form strong bonds with organic electronic materials (Wu, J. et al., 2010; Wu, Q. et al., 2010; Han et al., 2012; Sun et al., 2010; Hwang et al., 2012). Many OLEDs based on CVD graphene films have been reported (Sun et al., 2010; Hwang et al., 2012). Among them, Han et al. (2012) fabricated OLED devices based on CVD graphene films deposited on copper foils. They have observed that the OLEDs have extremely high performance compared with those based on conventional ITO (Han et al., 2012). In this case, graphene films transferred onto PET substrates that act as anode material were controlled in terms of electrical resistance, optical transmittance, and work function by random piling of multiple layers and acid doping (Han et al., 2012). The sheet resistance and work function varied from 189 to 89 Ohm/□ and 4.33 to 4.45 eV as a function of the number of layers. Han et al. (2012) used the devices at an applied voltage of 2 V and observed different luminances with respect to graphene as anode material. The luminance efficiency of devices incorporating graphene films is proportional to the number of graphene layers, the current density, and work function of graphene film anodes. Graphene film anodes doped with gold show higher luminance, which could be attributed to higher conductance and work function. Han et al. (2012) observed that the device using a four-layered graphene anode doped with nitric acid shows the highest luminous efficiency of 37.2 lm/W, whereas the device containing an ITO anode shows a luminous efficiency of 24.1 lm/W, indicating that indium degrades the efficiency of OLED devices. However, OLEDs incorporating graphene film anodes can thus emit homogeneous light with luminance intensity comparable with ITO and higher luminous efficiency. Therefore, graphene will allow new horizons in terms of designing hybrid architectures consisting of light-emitting semi-conductors.

1.4.9 Graphene as a Super-Capacitor/Ultra-Capacitors

Super-capacitors have attracted considerable attention as energy storage devices; they offer high power density, fast charge—discharge processes, and excellent cyclic stability (Kötz et al., 2000). Super-capacitors provide a good complement to Li—ion batteries in applications where both high energy (Li—ion batteries) and high power bursts are required, thus reducing the operational voltage dip at the load, extending energy efficiency and lifetime of the battery. Generally, super-capacitors were classified into two main types, electrical double-layer capacitors and pseuodo-capacitors. In electrochemical double-layer capacitors (EDLCs), a pure electrostatic attraction occurs between the ions accumulated at the electrode/electrolyte interface, with the electrode usually made of activated carbon. Carbon-based materials are widely used as electrode materials in double-layer capacitors because of their excellent physicochemical properties (Snook et al., 2011). In the pseudo-capacitors, electrons are additionally involved in quick Faradic reactions and are transferred to or from the valence bands of the redox cathode or anode reagent. Various transition metal oxides and conducting polymers have been used as electrodes in pseudo-capacitors because of their large surface area, π-conjugated length, and reversible redox processes. Likewise, graphene-based composite materials have extensive applications in the super-capacitors research field. Graphene-based nanocomposites with conducting polymers and metal oxides have been utilized for the applications in pseuodo-capacitors. The EDLCs capacitor consists of two porous carbon electrodes that are isolated from electrical contact by a porous separator (Pandolfo et al., 2006; Stoller et al., 2008). EDLCs are increasingly gaining attention because they fill a gap between batteries and ordinary capacitors. Super-capacitors find commercial applications for devices that need to store energy in the time range of $10^{-2}\,s < t < 10^2\,s$ (Kötz et al., 2000). Graphene is generally used instead of carbon for the development of novel electrode materials and is imperative for the design of high-performance ultra-capacitors that filter AC current in portable electronics equipment in EDLCs (Miller et al., 2010). Du et al (2010) have synthesized two kinds of functionalized graphene sheets by adopting the low-temperature thermal exfoliation method. The first kind of functionalized graphene offered specific capacitance value of 230 F/g, whereas the second kind of functionalized graphene sheet has the specific capacitance value of ~ 100 F/g. Wang, H. et al. (2009), Wang, X. et al. (2009), Wang, Y. et al. (2009), Yu, A. et al. (2010), and Yu, D. et al. (2010) synthesized 25-nm graphene sheets via a vacuum filtration method, realizing a capacitance of 135 F/g, which is sufficient mechanical strength for the flexible, applicable, and simplified/lightweight architecture in transparent electronics. Chen, D. et al. (2012), Chen, S. et al. (2012), Chen, X. et al. (2012), and Chen, Y. et al. (2012) prepared graphene-activated carbon composite by chemical activation method and demonstrated these for super-capacitor applications, with the specific capacitance of 122 F/g. Jaidev and Ramprabhu (2012) prepared poly (p-phenylene-diamine)-graphene nanocomposites and obtained the maximum specific capacitance value of 248 F/g at a specific current density of 2 A/g, proving the versatile ability of the prepared composite toward super-capacitor applications. Another trend in optimizing the geometry of graphene is the integration of CNT creates a good quality hybrid CNT/graphene composite material (Yu, A. et al., 2010; Yu, D. et al., 2010) that could be used in fabrication of a capacitor with an average capacitance of 120 F/g, which is considerably higher than those of vertically aligned and nonaligned CNT electrodes. Hybrid CNT/graphene composite is also used together with polyaniline (PANI), a conducting polymer, for the fabrication of capacitor with a high capacitance of 210 F/g at 0.3 A/g, and ~94% of this value (197 F/g) was maintained as the discharging current density was increased from 0.3 to 6 A/g (Wu, J. et al., 2010; Wu, Q. et al., 2010). Kim, B.J. et al. (2010) and Kim, K. et al. (2010) obtained a value of specific capacitance of 1118 F/g at a current density of 0.1 A/g, even though this figure dropped by 16% after 500 cycles. Li−ion batteries have high energy densities of ~180 Wh/kg and low power densities of ~1 kW/kg. Super-capacitors can deliver very high power densities of ~10 kW/kg with lower stored energy than batteries (~5 Wh/kg) (Conway, 1999). Recent graphene surface-enhanced lithium ion exchange cells seem to provide a solution for making an electrochemical energy storage device with both high energy density and power density (Jang et al., 2011). The approach was based on the exchange of lithium ions between the surfaces of two nanostructured electrodes, completely obviating the need for lithium intercalation or de-intercalation. In both electrodes, massive graphene surfaces in direct contact with liquid electrolyte are capable of rapidly and reversibly capturing lithium ions through surface adsorption and/or surface redox reaction. Energy density of 160 Wh/kg was obtained, which is 30-times higher than that (5 Wh/kg) of conventional symmetric super-capacitors and comparable with that of Li−ion batteries. They are also capable of delivering a power density of 100 kW/kg, which is 10-times higher than that (10 kW/kg) of super-capacitors and 100-times higher than that (1 kW/kg) of Li−ion batteries.

1.4.10 Spintronics

Since the discovery of the giant magnetoresistance (GMR) effect, extensive research has been devoted to finding new materials for application in spintronic devices. GMR devices (spin valves) basically consist of artificial thin film materials of alternate ferromagnetic and nonmagnetic layers. Resistance of the materials is at a minimum when the magnetic moments in ferromagnetic layers are aligned and maximum when they are anti-aligned. The major problem in spin valve structures is the interface scattering of spins between magnetic and nonmagnetic layers. To fabricate a clean (defect-free) interface, where there is no scattering, is a great technological challenge. Therefore, the ideal and most effective effort would be to search for a suitable material that intrinsically behaves as a spin valve, with two ferromagnetic edges separated by a nonmagnetic core. This property was finally realized in graphene, which is composed of a single atomic layer of carbon atoms arranged in a honeycomb lattice. A few layers of graphene are known as multilayer graphene, and many layers constitute the common material graphite. The gate-tunable, room temperature spin transport makes it an attractive material for spintronic applications. Graphene is special for spintronics for a number of reasons:

- Gate-tunable spin transport at room temperatures.
- Long spin diffusion lengths of ~4 microns at room temperature.
- Unusual Dirac band structure (like massless particles), which leads to predictions of unusual magnetic properties such as magnetized edges and half-metallic spin ordering in nanoribbons.
- Possibility of long spin lifetimes due to weak intrinsic spin−orbit and hyperfine interactions.
- Predictions of unusual gate-dependent magnetic and superconducting behavior in doped graphene.
- Large electron velocity that provides transport of spin polarized currents to exceptionally long distances.
- Extreme surface sensitivity that could be exploited for novel functionality including spin manipulation.
- Designs for spin-based computing in graphene, which can surpass Si CMOS for data-intensive applications.

1.4.10.1 Many Challenges and Opportunities Await for Spin and Magnetism in Graphene

McCreary et al. (2012) utilize pure spin currents to demonstrate that hydrogen adatoms and lattice vacancies generate magnetic moments in single-layer graphene (SLG). Pure spin currents are injected into graphene spin valve devices and clear signatures of magnetic moment formation emerge in the nonlocal spin transport signal as hydrogen adatoms or lattice vacancies are introduced in an ultrahigh vacuum environment. They have observed the effective exchange fields due to the spin–spin couplings, which are of interest for novel phenomena and spintronic functionality (Haugen et al., 2008; Semenov et al., 2007; Michetti et al., 2010; Qiao et al., 2010; Ray et al., 2014) in graphene. Swartz et al. (2013) measured and studied the nonlocal graphene tunneling spin valves of Mg-adatoms exposed to SLG and observed little variation of the spin relaxation times despite pronounced changes in the charge transport behavior. They found that the charge transport properties exhibit decreased mobility and decreased momentum scattering times. Han et al. (2011) investigated spin relaxation in graphene spin valves and observed strongly contrasting behavior of SLG and bilayer graphene. In SLG, the spin lifetime (τ_s) varies linearly with the momentum scattering time (τ_p), whereas the bilayer graphene shows the τ_s and τ_p exhibit an inverse dependence.

1.4.11 Integrated Circuits

For integrated circuits, graphene has high carrier mobility as well as low noise, allowing it to be used as the channel in a field effect transistor (FET). Single sheets of graphene are difficult to produce and even more difficult to make on an appropriate substrate (Chen et al., 2007). Ponomarenko et al. (2008) fabricated the smallest transistor so far using graphene that is one atom thick and 10 atoms wide. Wang, H. et al. (2009), Wang, X. et al. (2009), and Wang, Y. et al. (2009) fabricated an n-type transistor, meaning that both n-type and p-type graphene transistors had been created. A functional graphene integrated circuit was demonstrated—a complementary inverter consisting of one p-type and one n-type graphene transistor (Traversi et al., 2009). However, this inverter suffered from very low voltage gain. Lin et al. (2010) grew epitaxial graphene on SiC with quantity and quality suitable for mass production of integrated circuits. The first graphene-based integrated circuit was fabricated with a broadband radio mixer, and the circuit handled frequencies up to 10 GHz. Its performance was unaffected by temperatures up to 127°C (Lin et al., 2011).

1.4.12 Transistors

Outstanding challenges for graphene transistors include opening a sizeable and well-defined bandgap in graphene, making large-area graphene transistors that operate in the current saturation regime and fabricating GNRs with well-defined widths and clean edges. Graphene has semimetal linear energy dispersion, linear density of electronic states, and unique electronic structure substantially different from that of materials traditionally used in solid-state electronics. In the low energy range with respect to the Fermi level, the conduction and valence bands form conic shapes (referred to as "Dirac cones") and meet each other at the so-called Dirac points (Novoselov et al., 2005). One of the most important properties of graphene is a strong electric field effect that leads to an electrostatically tunable carrier density in the range of $n < 10^{14} \, cm^{-2}$ and it has high carrier mobilities for both electrons and holes (Chen et al., 2009; Farmer et al., 2009) (as high as $10^4 \, cm^2/Vs$ at room temperature) that attract a lot of attention to graphene as a possible material for future high-speed FET (Geim and Novoselov, 2007). The most often studied graphene FET structure is a back-gated configuration in which the graphene flake is contacted to form source and drain electrodes and the substrate acts as a back gate. By depositing a dielectric layer on top of such a device, one can achieve a top-gate configuration allowing both gate biases to control the charge concentration in the device channel. By synthesizing graphene on silicon carbide wafers (SiC) covering the whole SiC wafer, a large number of devices each having only one (top) gate can be fabricated (Klekachev et al., 2013).

A graphene metal oxide semi-conductor (MOS) device was among the breakthrough results reported by the Manchester group in 2004 (Novoselov et al., 2004). In this device, a 300-nm SiO_2 layer underneath the graphene served as a back-gate dielectric and a doped silicon substrate acted as the back-gate. Top-gated graphene MOSFETs have been made with graphene grown on metals such as nickel and copper (Kedzierski, 2009; Li, J. et al., 2009; Li, X. et al., 2009) and epitaxial graphene (Kedzierski, 2008; Lin et al., 2010; Moon et al., 2009); SiO_2, Al_2O_3, and HfO_2 have been used for the top-gate dielectric. The channels of these top-gated graphene transistors have been made using large-area graphene. Kim, B.J. et al. (2010) and Kim, K. et al. (2010) fabricated high-performance low-voltage flexible graphene FET with ion gel gate dielectrics on a flexible polymer substrate. This operation yielded a high on-current and low-voltage operation less than 3 V and showed a hole and electron mobility of 203 ± 57 and $91 \pm 50 \, cm^2/$ $(V \cdot s)$, respectively, at a drain bias of -1 V. Britnell et al. (2012) fabricated a bipolar FET that exploits the low density of states in graphene and its thickness of one atomic layer. These devices are based on graphene hetero-structures with atomically thin boron nitride or molybdenum disulfide acting as a vertical transport barrier and exhibiting switching ratios of ~50 and ~10,000, respectively, at room temperature. Such devices have potential for high-frequency operation and large-scale integration.

1.4.13 Ballistic Transistors

The availability of high mobility graphene up to room temperature makes ballistic transport in nanodevices achievable. The p−n−p transistors in the ballistic regime give access to Klein tunneling physics and allow the realization of devices exploiting the optics-like behavior of Dirac Fermions (DFs) as in the Veselago lens or the Fabry−Pérot cavity. Wilmart et al. (2014) propose a Klein tunneling transistor based on the geometrical optics of DFs. They considered a prismatic active region delimited by a triangular gate, where total internal reflection may occur, which leads to the tunable suppression of transistor transmission. Liang et al. (2007) also studied the performance projections for ballistic GNR FET and found that it behaves as high-mobility digital switches and has the potential to outperform the silicon MOSFET.

1.4.14 Radio Frequency Applications

The development of transparent radio frequency electronics has been limited, until recently, by the lack of suitable materials. Naturally thin and transparent graphene may lead to disruptive innovations in such applications. Wu et al. (2014) realized optically transparent broadband absorbers operating in the millimeter wave regime achieved by stacking graphene bearing quartz substrates on a ground plate. Broadband absorption is a result of mutually coupled Fabry−Pérot resonators represented by each graphene−quartz substrate. Millimeter wave reflectometer measurements of the stacked graphene−quartz absorbers demonstrated excellent broadband absorption of 90% with a 28% fractional bandwidth from 125 to 165 GHz bandwidth. This result suggests that the absorber operation could be extended to microwave and low-terahertz bands with negligible loss in performance. The transparent material could be used as a coating for car windows or buildings to stop radio waves from traveling through the structure and to improve secure wireless network environments.

1.4.14.1 Nano Antennas

Graphene material supports the propagation of tightly confined surface plasmon polaritons (SPP) waves. Due to their highly effective mode index, the SPP propagation speed can be up to two orders of magnitude below the electromagnetic (EM) wave propagation speed in vacuum. The main consequence is to reduce the resonant frequency of the antenna (Best et al., 2002). The work of Ryzhii et al. (2009) points to THz bands at short ranges, an unexplored range that allows wireless communications, thereby enabling graphene-enabled wireless communications (GWC) (Yang et al., 2010). Recently, Brian (2014) approached a graphene-based plasmonic nanoantenna (GPN) that can operate efficiently at millimeter radio wavelengths. Unlike plasmonic antennas based on noble metals, these would allow smart dust to operate at frequencies at least 100- to 1000-times smaller than is possible using a conventional metallic antenna. When the EM wave directed onto a graphene surface perpendicular to that surface, it excited the electrons in the graphene into oscillations. When resonance occurs, then the coupling of the SPP/external EM waves increases greatly, resulting in efficient transfer of energy between the two. In this case, a phased array antenna 100 μm in diameter could produce 300 GHz beams only a few degrees in diameter, instead of the 180-degree radiation from a conventional metal antenna of that size. Brian predicted that there are other applications for such small antennas. This could also be used to make practical terabit-per-second wireless networks for smart phones and computers while avoiding being crippled by the power loss associated with the propagation of sub-THz radio waves through the atmosphere.

1.4.15 Sound Transducer
Tian et al. (2012) demonstrated that SLG emits sound. It significantly produces a flat frequency response in the ultrasound range from 20 to 50 kHz and produces 95 dB at a distance of 5 cm with a sound frequency of 20 kHz. This device could reduce the conventional acoustic devices with a large size to nanoscale. It has the potential to be widely used in speakers, buzzers, earphones, ultrasonic transducer, and others. Graphene has provided relatively good frequency response in audio speakers, and its light weight may make it suitable for microphones as well.

1.4.16 Graphene as Sensor
Sensors are widely studied and used in our daily life, and their applications are increasing in many fields, such as electrochemical, biological, and clinical diagnosis and environment detections like industry

(pollutant) and research institutes (radiation measurements). The exceptional properties of graphene allow for development of sensors of various types (Hill et al., 2011). Each atom in the graphene sheet interacts directly with the sensing environment; the electronic properties of graphene can be modified by this interaction.

1.4.16.1 Electrochemical Sensor

Graphene and graphene oxide nanocomposite-based materials have received considerable attention for the fabrication of several electrochemical sensors due to their low cost, high catalytic ability, and good stability (Shan et al., 2009a,b; Zhou et al., 2009; Kang et al., 2009). Fan et al. (2011) and Ameer et al. (2012) prepared graphene/PANI nanocomposite and utilized it for the determination of 4-aminophenol via DPV technique and highly sensitive detection of hydrazine sensor, respectively. Du et al. (2013) used graphene-modified carbon fiber electrodes for the nonenzymatic electrochemical uric acid sensors within the range of 0.194–49.68 μM within the detection value 0.132 μM. Carbon nanotubes–graphene hybrid (SWCNT-GNS) electrode was used by Chen, D. et al. (2012), Chen, S. et al. (2012), Chen, X. et al. (2012), and Chen, Y. et al. (2012) for the detection of acetaminophen by DPV, showing a high performance range from 0.05 to 84.5 μM with a low detection limit of 38 nM.

1.4.16.2 Gas Sensors

Graphene has no dangling bonds on its surface, so gaseous molecules cannot be readily adsorbed onto its surface. By coating with a thin layer of certain polymers on graphene, the sensitivity of chemical gas sensors can be enhanced. The absorption of gas molecules acts as a concentrator that introduces a local change in electrical resistance. This effect also occurs in other materials, but graphene is superior due to its high electrical conductivity and low noise, making even small changes in resistance detectable (Schedin et al., 2007).

1.4.16.3 Biosensors

Graphene-based biosensors were extensively studied because of the large specific area and good electrical, thermal, and biocompatibility properties of graphene. Polypyrrole–graphene–glucose oxidase-based biosensor was fabricated with graphene sheets that were covalently attached to glucose oxidase, and the resulting modified electrode has been used for the determination of glucose. Liu, J. et al., (2012) and Liu, M. et al., (2012) fabricated a phenylethynyl ferrocene/graphene nanocomposite-based dopamine biosensor for the sensitive and selective detection of dopamine in serum and urine. Shan et al. (2009a,b) reported the first graphene-based glucose biosensor with a graphene/polyethylenimine-functionalized ionic liquid nanocomposites-modified electrode that exhibits glucose response (2−14 mM, R¼ 0.994) and high stability. Shan et al. (2009a,b) also reported a graphene/AuNPs/chitosan composite film-based biosensor that exhibited good electrocatalytical activity toward H_2O_2 and O_2. Thionine−graphene nanocomposite was developed by Zhu et al. (2012) for the electrochemical biosensing of DNA; it possesses good selectivity and a wide range of $1.0 \times 10^{-12}-1.0 \times 10^{-7}$ M and with a very low detection limit of 1.26×10^{-13} M. Li, J. et al. (2009) and Li, X. et al. (2009) reported that the Nafion−graphene composite film-based electrochemical sensors not only exhibit improved sensitivity for the metal ion (Pb^{2+} and Cd^{2+}) detections but also alleviate the interferences due to the synergistic effect of GNSs and Nafion.

1.4.17 Composite Materials

It has already been established that the addition of CNTs to polymer matrices improve the mechanical, electrical, and thermal properties (Coleman et al., 2006). Now, the challenge is for SLG to be used in large quantity as an inexpensive and feasible substitute for CNTs. It was demonstrated that the incorporation of well-dispersed graphene sheets into polymers nanocomposites/graphene−nanometals−polymer hybrids at an extraordinarily low filler content results in remarkable impact not only on the mechanical properties of the polymer (Ramanathan et al., 2008) but also on engineered electrical and thermal conductivity, fundamental characteristics of avionic/space, and homeland security applications. Different applications of the graphene composite and inks are as follows:

- Fuel tank coatings
- Polymers with EMI or RFI shielding capabilities
- Automotive composites
- Electronic enclosures
- Photonic composites
- Aerospace composite and EMI shielding
- Heat dissipater in electrical appliances
- Sporting goods

Coating and placement of graphene inks can be achieved via spin coating (Eda et al., 2008), spray coating (Blake et al., 2008), and screen printing (Qian et al., 2009). Inkjet printing can be directly integrated in processing of electronic and optoelectronic devices (Torrisi, 2012). Graphene ceramic composites also possess further advantages compared with their CNT counterparts, such as the lower cost and commercial availability of graphene and the less stringent processing conditions. Graphene/ceramic composites may find applications in friction and wear-related fields, such as engine components, bearings, and cutting tools for metal working operations. These composites have already been used in sliding contacts and as a solid lubricant (Gonzalez-Julian et al., 2011). Photonic polymer–graphene composite materials are also very useful in optical communications. Some researchers are trying to develop a new class of polymer-based optoelectronic devices embedding the optical and electronic functionalities of graphene, 2D inorganic-layered materials, and their hybrid heterostructures. These devices will combine the fabrication advantages of polymer photonics with tunable-active and tunable-passive optical properties of such materials. Such devices are expected to find a wide range of applications not only in optical communications but also in biomedical instruments, chemical analysis, time-resolved spectroscopy, electro-optical sampling, microscopy, and surgery.

1.4.18 Liquid Crystal Displays

Graphene has high electrical conductivity and excellent transmittance at terahertz frequencies, so Wu, M.S. et al. (2013), Wu, S. et al. (2013), and Wu, Y. et al. (2013) demonstrated a liquid crystal-based terahertz phase shifter with this graphene. They observed in a 50-μm liquid crystal cell that the maximum phase shift is 10.8° with saturation voltage of 5 V. This phase shifter provides continuous tunability, fully electrical controllability, and low DC voltage operation that depends on the number of graphene layers. Blake et al. (2008) also demonstrated liquid crystal devices with electrodes made of graphene that show excellent performance with a high contrast ratio that can be comparable with conventionally used metal oxides in terms of low resistivity, high transparency, and chemical stability.

1.4.19 Graphene Quantum Dots

All dimensions of graphene quantum dots (GQDs) are <10 nm. Their size and edge crystallography govern their electrical, magnetic, optical, and chemical properties. GQDs can be produced via graphite nanotomy (Mohanty et al., 2012) or via bottom-up, solution-based routes (Cai et al., 2010). GQDs with controlled structures can be incorporated into applications in electronics, optoelectronics, and EMs. Quantum confinement can be created by changing the width of GNRs at selected points along the ribbon (Ponomarenko et al., 2008).

1.4.20 Frequency Multiplier
In 2009, researchers built an experimental graphene frequency multiplier that takes an incoming signal of a certain frequency and outputs a signal at a multiple of that frequency (Wang, H. et al., 2009; Wang, X. et al., 2009; Wang, Y. et al., 2009; Cricchio et al., 2009).

1.4.21 Optical Modulator
When the Fermi level of graphene is tuned, its optical absorption can be changed. In 2011, researchers reported the first graphene-based optical modulator. Operating at 1.2 GHz without a temperature controller, this modulator has a broad bandwidth (from 1.3 to 1.6 µm) and small footprint ($\sim$25 µm^2) (Liu et al., 2011).

1.4.22 Infrared Light Detection
Graphene reacts to the infrared spectrum at room temperature, albeit with sensitivity 100- to 1000-times too low for practical applications. However, two graphene layers separated by an insulator allowed an electric field produced by holes left by photo-freed electrons in one layer to affect a current running through the other layer. The process produces little heat, making it suitable for use in night-vision optics. The sandwich is thin enough to be integrated in handheld devices, eyeglass-mounted computers, and even contact lenses.

1.4.23 Graphene Photodetectors
Photodetectors measure photon flux or optical power by converting the absorbed photon energy into electrical current. A wide range of devices (Saleh and Teich, 2007) are available, such as remote controls, televisions, and DVD players. With an internal photo-effect, the absorption of photons results in carriers excited from the valence to the conduction band, outputting an electric current. The spectral bandwidth is typically limited by the absorption (Saleh and Teich, 2007). Graphene absorbs from the UV to THz (Dawlaty et al., 2008; Wright et al., 2009). As a result, graphene-based photo-detectors could work

over a much broader wavelength range. The response time is ruled by the carrier mobility, and graphene has great mobility, so it can be ultrafast (Saleh and Teich, 2007). Graphene's suitability for high-speed photodetection was demonstrated in a communications link operating at 10 Gbit/s (Mueller et al., 2010).

1.4.24 Piezoelectricity
Density functional theory simulations predict that depositing certain adatoms on graphene can render it piezoelectricity responsive to an electric field applied in the out-of-plane direction. This type of locally engineered piezoelectricity is similar in magnitude to that of bulk piezoelectric materials and makes graphene a candidate for control and sensing in nanoscale devices (Ong et al., 2012).

1.4.25 Graphene as Purification of Water
Graphene sheets perforated by small holes were first explored as potential candidates for water filtration by researchers at MIT. Holes with a diameter of 1 nanometer (a billionth of a meter) are big enough to let water molecules sift through, but they are small enough to stop any undesired chemicals. Han et al. (2013) fabricated ultrathin ($\sim$22–53 nm thick) graphene nanofiltration membranes (uGNMs) on microporous substrates that were used for efficient water purification. The performance of the uGNMs for water treatment was evaluated on a dead-end filtration device, and the pure water flux of uGNMs was high (21.8 L/m^2/h/bar). The uGNMs showed high retention ($>$99%) for organic dyes and moderate retention ($\sim$20–60%) for ion salts.

1.5 CONCLUSION AND PERSPECTIVES OF GRAPHENE

Graphene, a 2D monoatomic thick building block of a carbon allotrope, has emerged as an exotic material of the 21st century and has received worldwide attention because of its exceptional charge transport, thermal, optical, and mechanical properties. Graphene and its derivatives are being studied in nearly every field of science and engineering. Graphene is one of the most promising and versatile enabling nanotechnologies addressing *"secure, clean, and efficient energy."* Recent progress has shown that graphene-based materials can have a profound impact on electronic and optoelectronic devices, chemical sensors, nanocomposites, and energy storage. Graphene will bring disruptive solutions to the current industrial challenges related to energy generation and storage applications, first in nano-enhanced products, and then in radically new nano-enabled products. Graphene-based systems for energy production (photovoltaics, fuel cells), energy storage (super-capacitors, batteries), and hydrogen storage will be developed via relevant proof-of-concept demonstrators that will progress toward the targeted technology readiness levels required for industrial uptake.

1.6 THE PRESENT CHALLENGES AND FUTURE RESEARCH IN GRAPHENE NANOMATERIALS

The exceptional properties of graphene, including electrical, thermal, mechanical, optical, and long electron mean free paths, make it compelling for various engineering applications. Huge efforts have been devoted to the fundamental physics and chemistry of graphene. Properties such as room temperature quantum Hall effect, highest charge transport, and thermal conductivity originated from graphene's 2D structures have not been observed from most conventional 3D materials. Several research publications in the past 10 years signify the importance of graphene and that it might surpass silicon research in the development of microelectronics. Although silicon-based research is at a mature stage to overcome technological barriers, graphene is being extensively investigated because it holds the future for microscale to nanoscale electronics. The inherent semi-metal characteristic of graphene has been modified to realize the applications in transistors. Much effort has been devoted to modification of a graphene band gap, allowing applications in electronic devices. Graphene as a new material still faces many challenges ranging from synthesis and characterization to the final device fabrication. The exceptional properties were observed in the defect-free pristine graphene prepared by graphite exfoliation using the scotch tape method, which is not appropriate for any large-scale device manufacturing. Alternative methods have progressed to CVD and epitaxially grown single-layer, bilayer, and few-layer graphene on different substrates and simple transfer of the graphene layer for subsequent device fabrication. These breakthroughs offer novel and exciting opportunities for semi-conductor industries. The advanced deposition technique of the single layer and bilayer made it possible to fabricate large-area devices; however, creating band gap in a controlled and practical manner is still challenging for application in logic devices. Several methods aiming at tuning the substrate properties and nanoribbon dimension have been proposed to introduce a tunable band gap essential for nanoelectronics. This energy band gap can be achieved through quantum confinement, bilayer graphene, and chemical functionalization. The former, quasi-1D GNR has been considered with either edge localization or Coulomb blockade effects. GNR with appropriate dimensions (i.e., 10 nm) is expected to provide right band gap for efficient FET devices. Therefore, the band gap opened graphene based on GNR has been encouraged for future practical nanoelectronic devices comparable with complementary MOS circuits.

REFERENCES

Allen, M.J., et al., 2010. Honeycomb carbon: a review of graphene. Chem. Rev. 110, 132–145.

Ameer, S., et al., 2012. Hydrazine chemical sensing by modified electrode based on in situ electro-chemically synthesized polyaniline/graphene composite thin film. Sens. Actuators B 173, 177–183.

Arco, L.G.D., et al., 2010. Continuous, highly flexible, and transparent graphene films by chemical vapor deposition for organic photovoltaics. ACS Nano 4, 2865–2873.

Ata, M.S., et al., 2012. Electrophoretic deposition of graphene, carbon nanotubes and composites using aluminon as charging and film forming agent. Colloids Surf. A Physicochem. Eng. Asp. 398, 9–16.

Atabaki, M.M., et al., 2013. Graphene composites as anode materials in lithium–ion batteries. Electron. Mater. Lett. 9, 33–153.

Ataca, C., et al., 2008. High-capacity hydrogen storage by metallized graphene. Appl. Phys. Lett. 93, 1–3, 043123.

Balog, R., et al., 2009. Atomic hydrogen adsorbate structures on graphene. J. Am. Chem. Soc. 131, 8744–8745.

Bae, S., et al., 2010. Roll-to-roll production of 30-inch graphene films for transparent electrodes. Nat. Nanotechnol. 5, 574–578.

Bae, S., et al., 2012. Towards industrial applications of graphene electrodes. Phys. Scr. T146 (014024), 1–8.

Basu, S., et al., 2014. Graphene-based electrodes for enhanced organic thin film transistors based on pentacene. Phys. Chem. Chem. Phys. 16, 16701–16710.

Bell, J.S., et al., 1998. Platinum-macrocycle co-catalysts for the electrochemical oxidation of methanol. Electrochim. Acta 43, 3645–3655.

Bendikov, M., et al., 2004. Tetrathiafulvalenes, oligoacenenes, and their buckminsterfullerene derivatives: the brick and mortar of organic electronics. Chem. Rev. 104, 4891–4946.

Best, S.R., et al., 2002. IEEE Antennas Wirel. Propag. Lett. 1, 112–115.

Blake, P., et al., 2008. Graphene-based liquid crystal device. Nano Lett. 8, 1704–1708.

Bonaccorso, F., et al., 2010. Graphene photonics and optoelectronics. Nat. Photon. 4, 611–622.

Brian, D., 2014. Graphene-based nano-antennas may enable cooperating smart dust swarms. <http://www.gizmag.com/graphene-nano-antennas-smart-dust-swarm/30373/>.

Britnell, L., et al., 2012. Field-effect tunneling transistor based on vertical graphene heterostructures. Science 335, 947–950.

Brownson, D.A.C., et al., 2011. An overview of graphene in energy production and storage applications. J. Power Sources 196, 4873–4885.

Bundgaard, E., et al., 2010. Low band gap polymers for roll-to-roll coated polymer solar cells. Macromolecules 43, 8115–8120.

Burress, J.W., et al., 2010. Graphene oxide framework materials: theoretical predictions and experimental results. Chem. Int. Ed. 49, 8902–8904.

Cai, J., et al., 2010. Atomically precise bottom-up fabrication of graphene nanoribbons. Nature 466 (7305), 470–473.

Chandra, S., et al., 2010. A novel synthesis of graphene by dichromate oxidation. Mater. Sci. Eng. B 167, 133–136.

Charrier, A., et al., 2002. Solid-state decomposition of silicon carbide for growing ultra-thin heteroepitaxial graphite films. J. Appl. Phys. 92 (5), 2479–2484.

Chen, J., et al., 2007. Printed graphene circuits. Adv. Mater. 19 (21), 3623–3627.

Chen, F., et al., 2009. Dielectric screening enhanced performance in graphene FET. Nano Lett. 9, 2571–2574.

Chen, Y., et al., 2010. Electrophoretic deposition of graphene nanosheets on nickel foams for electrochemical capacitors. J. Power Sources 195, 3031–3035.

Chen, F., et al., 2011. Functional materials for rechargeable batteries. Adv. Mater. 23, 1695–1715.

Chen, D., et al., 2012. Reversible lithium–ion storage in silver-treated nanoscale hollow porous silicon particles. Angew. Chem. Int. Ed. 51, 2409–2413.

Chen, S., et al., 2012. Thermal conductivity of isotopically modified graphene. Nat. Mater. 11, 203–207.

Chen, X., et al., 2012. A high performance electrochemical sensor for acetaminophen based on single-walled carbon nanotube–graphene nanosheet hybrid films. Sens. Actuators B 161, 648–654.

Chen, Y., et al., 2012. High-performance supercapacitors based on a graphene-activated carbon composite prepared by chemical activation. RSC Adv. 2 (2012), 7747–7753.

Choi, D., et al., 2010. Li-ion batteries from $LiFePO_4$ cathode and anatase/graphene composite anode for stationary energy storage. Electrochem. Commun. 12, 378–381.

Coleman, J.N., et al., 2006. Mechanical reinforcement of polymers using carbon nanotubes. Adv. Mater. 18, 689–706.

Conway, B.E., 1999. Electrochemical Supercapacitors: Scientific Fundamentals and Technological Applications. Springer, New York, NY.

Coraux, J., et al., 2008. Structural coherency of graphene on Ir(1 1 1). Nano Lett. 8 (2), 565–570.

Cricchio, D., et al., 2009. A paradigm of fullerene. J. Phys. B 42 (8), 1–7, 085404.

Dawlaty, J.M., et al., 2008. Measurement of the optical absorption spectra of epitaxial graphene from terahertz to visible. Appl. Phys. Lett. 93, 1–3, 131905.

Dimitrakakis, G.K., et al., 2008. Pillared graphene: a new 3-D network nanostructure for enhanced hydrogen storage. Nano Lett. 8, 3166–3170.

Du, A., et al., 2010. Multifunctional porous graphene for nanoelectronics and hydrogen storage: new properties revealed by first principle calculations. J. Am. Chem. Soc. 132, 2876–2877.

Du, H.Y., et al., 2012. Graphene nanosheet-CNT hybrid nanostructure electrode for a proton exchange membrane fuel cell. Int. J. Hydrogen Energy 37, 18989–18995.

Du, J., et al., 2013. Nonenzymatic uric acid electrochemical sensor based on graphene-modified carbon fiber electrode. Colloids Surf. A Physicochem. Eng. Asp. 419, 94–99.

Durgun, E., et al., 2008. Functionalization of carbon-based nanostructures with light transition-metal atoms for hydrogen storage. Phys. Rev. B 77, 1–9, 085405.

Eda, G., et al., 2008. Large-area ultrathin films of reduced graphene oxide as a transparent and flexible electronic material. Nat. Nanotech. 3, 270–274.

Enachescu, M., et al., 1999. Integration of point-contact microscopy and atomic-force microscopy: application to characterization of graphite/Pt(1 1 1). Phys. Rev. B 60 (24), 16913–16919.

Enoki, T., et al., 2005. Magnetic nanographite: an approach to molecular magnetism. J. Mater. Chem. 15, 3999–4002.

Fan, Y., et al., 2011. Graphene–polyaniline composite film modified electrode for voltammetric determination of 4-aminophenol. Sens. Actuators B 157, 669–674.

Farmer, D.B., et al., 2009. Utilization of a buffered dielectric to achieve high field-effect carrier mobility in graphene transistors. Nano Lett. 9, 4474–4478.

Ferro, Y., et al., 2008. Stability and magnetism of hydrogen dimers on graphene. Phys. Rev. B 78, 1–8, 085417.

Florescu, L.G., et al., 2013. The influence of dilution gases on multilayer graphene formation in laser pyrolysis. Appl. Surf. Sci. 278, 313–316.

Fujita, T.K.W., et al., 2005. Novel structures of carbon layers on a Pt(1 1 1) surface. Surf. Interface Anal. 37 (2), 120–123.

Geim, A.K., 2009. Graphene: status and prospects. Science 324, 1530–1534.

Geim, A.K., Novoselov, K.S., 2007. The rise of graphene. Nat. Mater. 6, 183–191.

Golabi, S.M., et al., 2002. Electrocatalytic oxidation of methanol on electrodes modified by platinum microparticles dispersed into poly(o-phenylenediamine) film. J. Electroanal. Chem. 521, 161–167.

Gonzalez-Julian, J., et al., 2011. Enhanced tribological performance of silicon nitride-based materials by adding carbon nanotubes. J. Am. Ceram. Soc. 94, 2542–2548.

Gratzel, M., 2007. Photovoltaic and photoelectrochemical conversion of solar energy. Philos. Trans. R. Soc. A: Math. Phys. Eng. Sci. 365 (1853), 993–1005.

Guisinger, N.P., et al., 2009. Exposure of epitaxial graphene on SiC(0001) to atomic hydrogen. Nano Lett. 9, 1462–1466.

Guo, G.F., et al., 2012. Electrochemical hydrogen storage of the graphene sheets prepared by DC arc-discharge method. Surf. Coat. Technol. 228, S120–S125.

Han, W., et al., 2011. Spin relaxation in single-layer and bilayer graphene. Phy. Rev. Lett. 107, 1–4, 047207.

Han, T.H., et al., 2012. Extremely efficient flexible organic light-emitting diodes with modified graphene anode. Nat. Photonics 6, 105–110.

Han, Y., et al., 2013. Ultrathin graphene nano-filtration membrane for water purification. Adv. Funct. Mater. 23, 3693–3700.

Haugen, H., et al., 2008. Spin transport in proximity-induced ferromagnetic graphene. Phys. Rev. B 77, 1–8, 115406.

Hey, Y.S., et al., 2011. A novel bath lily-like graphene sheet-wrapped nano-Si composite as a high performance anode material for Li–ion batteries. RSC Adv. 1 (2011), 958–960.

Hill, E.W., et al., 2011. Graphene sensors. IEEE Sens. J. 11, 3161–3170.

Hong, B.H., 2011 "Synthesis and applications of graphene for flexible electronics" Graphene 2011 at Imagine nano 2011.

Hou, J., et al., 2013. A new method for fabrication of graphene/polyaniline nanocomplex modified microbial fuel cell anodes. J. Power Sources 224, 139–144.

Hu, C.C., et al., 1999. Voltammetric investigation of platinum oxides. I. Effects of ageing on their formation/reduction behavior as well as catalytic activities for methanol oxidation. Electrochim. Acta 44, 2727–2738.

Hwang, J., et al., 2012. Multilayered graphene anode for blue phosphorescent organic light emitting diodes. Appl. Phys. Lett. 100, 1–4, 133304.

Iijima, S., 1991. Helical microtubules of graphitic carbon. Nature 354, 56–58.

Jaidev, Ramprabhu, S., 2012. Poly(p-phenylenediamine)/graphene nanocomposites for supercapacitor applications. J. Mater. Chem. 22, 18775–18783.

Jang, B.Z., et al., 2011. Graphene surface-enabled lithium ion-exchanging cells: next-generation high-power energy storage devices. Nano Lett. 11 (9), 3785–3791.

Ju, S., 2010. Synthesis of Si/graphene composite as an anode material for lithium-ion battery. In: Meeting Abstracts of the 15th International Meeting on Lithium Batteries, ECS, vol. 78 (p. 5878).

Kakaei, K., et al., 2013. A new method for manufacturing graphene and electrochemical characteristic of graphene-supported Pt nanoparticles in methanol oxidation. J. Power Sources 225, 356–363.

Kang, X.H., et al., 2009. Glucose oxidase–graphene–chitosan modified electrode for direct electrochemistry and glucose sensing. Biosens. Bioelectron. 25, 901–905.

Kedzierski, J., 2008. Epitaxial graphene transistors on SiC substrates. IEEE Trans. Electron Devices 55, 2078–2085.

Kedzierski, J., 2009. Graphene-on-insulator transistors made using C on Ni chemical-vapor deposition. IEEE Electron Device Lett. 30, 745–747.

Kim, G., et al., 2008. Optimization of metal dispersion in doped graphitic materials for hydrogen storage. Phys. Rev. B 78, 1–5, 085408.

Kim, B.J., et al., 2010. High-performance flexible graphene field effect transistors with ion gel gate dielectrics. Nano Lett. 10, 3464–3466.

Kim, K., et al., 2010. Influence of multi-walled carbon nanotubes on the electrochemical performance of graphene nanocomposites for supercapacitor electrodes. Electrochim. Acta 56 (3), 1629–1635.

Klekachev, A.V., et al., 2013. Graphene transistors and photodetectors. Electrochem. Soc. Interface Spring, 63–68.

Kötz, R., et al., 2000. Principles and applications of electrochemical capacitors. Electrochim. Acta 45, 2483–2498.

Kou, R., et al., 2009. Enhanced activity and stability of Pt catalysts on functionalized graphene sheets for electrocatalytic oxygen reduction. Electrochem. Commun. 11, 954–957.

Kroto, H.W., et al., 1985. C60: Buckminsterfullerene. Nature 318, 162–163.

Kumar, A., et al., 2013. Nitrogen-doped graphene by microwave plasma chemical vapor deposition. Thin Solid Films 528, 269–273.

Land, T.A., et al., 1992. STM investigation of single layer graphite structures produced on Pt(1 1 1) by hydrocarbon decomposition. Surf. Sci. 264 (3), 261–270.

Lee, H., et al., 2010. Calcium-decorated graphene-based nanostructures for hydrogen storage. Nano Lett. 10, 793–798.

Lee, Y., et al., 2013. Graphene based transparent conductivity films. NANO: Brief Rep. Rev. 8, 1–16, 1330001.

Li, J., et al., 2009. High-sensitivity determination of lead and cadmium based on the Nafion–graphene composite film. Anal. Chim. Acta 649, 196–201.

Li, X., et al., 2009. Large-area synthesis of high-quality and uniform graphene films on copper foils. Science 324, 1312–1314.

Liang, G., et al., 2007. Performance projections for ballistic graphene nanoribbon field-effect transistors. IEEE Trans. Electron Devices 54, 677–682.

Liang, M., et al., 2009. Graphene-based electrode materials for rechargeable lithium batteries. J. Mater. Chem. 19 (33), 5871–5878.

Lin, Y.-M., et al., 2010. 100-GHz transistors from wafer-scale epitaxial graphene. Science 327 (5966), 662–662.

Lin, Y.-M., et al., 2011. Wafer-scale graphene integrated circuit. Science 332 (6035), 1294–1297.

Liu, Z., et al., 2008. Organic photovoltaic devices based on a novel acceptor material: graphene. Adv. Mater. 20 (20), 3924–3930.

Liu, C., et al., 2010. Membraneless enzymatic biofuel cells based on graphene nanosheets. Biosens. Bioelectron. 25, 1829–1833.

Liu, Y., et al., 2010. Titanium-decorated graphene for high-capacity hydrogen storage studied by density functional simulations. J. Phys. Condens. Matter. 22, 1–5, 445301.

Liu, Z., et al., 2010. Improving photovoltaic properties by incorporating both SPF graphene and functionalized multi-walled carbon nanotubes. Sol. Energy Mater. Sol. Cells 94 (12), 2148–2153.

Liu, M., et al., 2011. A graphene-based broadband optical modulator. Nature 474 (7349), 64–67.

Liu, J., et al., 2012. Graphene/carbon cloth anode for high-performance mediatorless microbial fuel cells. Bioresour. Technol. 114, 275–280.

Liu, M., et al., 2012. Highly sensitive and selective dopamine biosensor based on a phenylethynyl ferrocene/graphene nanocomposite modified electrode. Analyst 137, 4577–4583.

Mai, Y.J., et al., 2011. CuO/graphene composite as anode materials for lithium–ion batteries. Electrochim. Acta 56, 2306–2311.

McCreary, K.M., et al., 2012. Magnetic moment formation in graphene detected by scattering of pure spin currents. Phy. Rev. Lett. 109, 1–5, 186604.

Michetti, P., et al., 2010. Electric field control of spin rotation in bilayer graphene. Nano Lett. 10, 4463–4469.

Miller, J.R., et al., 2010. Graphene double-layer capacitor with ac line-filtering performance. Science 329 (5999), 1637–1639.

Mohanty, N., et al., 2012. Nanotomy based production of transferrable and dispersible graphene-nanostructures of controlled shape and size. Nat. Commun. 3 (5), 844–848.

Moon, J.S., et al., 2009. Epitaxial-graphene RF field-effect transistors on Si-face 6H–SiC substrates. IEEE Electron Device Lett. 30, 650–652.

Mueller, T., et al., 2010. Graphene photodetectors for high-speed optical communications. Nat. Photonics 4, 297–301.

Nagashima, A., et al., 1994. Change in the electronic states of graphite overlayers depending on thickness. Phys. Rev. B 50 (7), 4756–4763.

Neto, A.H.C., et al., 2009. The electronic properties of graphene. Rev. Mod. Phys. 81, 109–162.

Novoselov, K.S., et al., 2004. Electric field effect in automically thin carbon films. Science 306, 666–669.

Novoselov, K.S., et al., 2005. Two-dimensional gas of massless Dirac Fermions in graphene. Nature 438, 197–200.

Ong, M.T., et al., 2012. Engineered piezoelectricity in graphene. ACS Nano 6 (2), 1387–1394.

Pandolfo, A.G., et al., 2006. Carbon properties and their role in supercapacitors. J. Power Sources 157 (1), 11–27.

Papagno, L., et al., 1984. Determination of graphitic carbon structure adsorbed on Ni(110) by surface extended energyloss fine structure analysis. Phys. Rev. B 29 (3), 1483–1486.

Parambhath, V.B., et al., 2011. Investigation of spillover mechanism in palladium decorated hydrogen exfoliated functionalized graphene. J. Phys. Chem. C 115, 15679–15685.

Park, H., et al., 2010. Doped graphene electrodes for organic solar cells. Nanotechnology 21, 505204.

Patchkovskii, S., et al., 2005. Graphene nanostructures as tunable storage media for molecular hydrogen. PNAS 102, 10439–10444.

Ponomarenko, L.A., et al., 2008. Chaotic Dirac billiard in graphene quantum dots. Science 320 (5874), 356–358.

Qian, M., et al., 2009. Electron field emission from screen-printed graphene films. Nanotechnology 20, 1–6, 425702.

Qiao, Z., et al., 2010. Quantum anomalous Hall effect in graphene from Rashba and exchange effects. Phys. Rev. B 82, 1–4, 161414.

Qu, J., et al., 2004. Preparation of hybrid thin film modified carbon nanotubes on glassy carbon electrode and its electrocatalysis for oxygen reduction. Chem. Commun. 1, 34–35.

Ramachandran, R., et al., 2013. Recent trends in graphene based electrode materials for energy storage device and sensor applications. Int. J. Electrochem. Soc. 8, 11680–11694.

Ramanathan, T., et al., 2008. Functionalized graphene sheets for polymer nanocomposites. Nat. Nanotech. 3, 327–331.

Rao, C.N., et al., 2009. Graphene: the new two-dimensional nanomaterial. Angew. Chem. Int. Ed. 48, 7752–7777.

Ray, S.C., et al., 2014. Graphene supported graphone/graphane bilayer nanostructure material for spintronics. Sci. Rep. 4, 1–7, 3862.

Ryoko, O., et al., 2000. Electronic states of monolayer graphene on Pt(755) and TiC(755). Tanso 195, 400–404.

Ryzhii, V., 2009. Feasibility of terahertz lasing in optically pumped epitaxial multiple graphene layer structures. J. Appl. Phys. 106, 1–6, 084507.

Saleh, B.E.A., Teich, M.C., 2007. Fundamentals of Photonics (Chapter 18). Wiley, New York, pp. 784–803.

Schedin, F., et al., 2007. Detection of individual gas molecules adsorbed on graphene. Nat. Mater. 6, 652–655.

Semenov, Y.G., et al., 2007. Spin field effect transistor with a graphene channel. Appl. Phys. Lett. 91, 1–3, 153105.

Shan, C.S., et al., 2009a. Graphene/AuNPs/chitosan nanocomposites film for glucose biosensing. Biosens. Bioelectron. 2009 (25), 1070–1074.

Shan, C.S., et al., 2009b. Direct electrochemistry of glucose oxidase and biosensing for glucose based on graphene. Anal. Chem. 81, 2378–2382.

Singh, V., et al., 2011. Graphene based materials: past, present and future. Prog. Mater. Sci. 56, 1178–1271.

Snook, G.A., et al., 2011. Conducting-polymer-based supercapacitor devices and electrodes. J. Power Sources 196, 1–12.

Sobkowski, J., et al., 1985. Influence of tin on the oxidation of methanol on a platinum electrode. J. Electroanal. Chem. 196, 145–156.

Sofo, J.O., et al., 2007. Graphane: a two-dimensional hydrocarbon. Phys. Rev. B 75, 1–4, 153401.

Soldano, C., et al., 2010. Production, properties and potential of graphene. Carbon 48, 2127–2150.

Stoller, M.D., et al., 2008. Graphene-based ultracapacitors. Nano Lett. 8 (10), 3498–3502.

Sun, T., et al., 2010. Multilayered graphene used as anode of organic light emitting devices. Appl. Phys. Lett. 96, 13301–13303.

Sutter, P.W., et al., 2008. Epitaxial graphene on ruthenium. Nat. Mater. 7 (5), 406–411.

Swartz, A.G., et al., 2013. Effect of in situ deposition of Mg adatoms on spin relaxation in graphene. Phy. Rev. B 87, 1–4, 075455.

Tanaka, T.I.A., et al., 2003. Heteroepitaxial system of h-BN/monolayer graphene on Ni(1 1 1). Surf. Rev. Lett. 10 (4), 697–703.

Tapia, A., et al., 2011. Potassium influence in the adsorption of hydrogen on graphene: a density functional theory study. Comput. Mat. Sci. 50, 2427–2432.

Terai, M., et al., 1998. Electronic states of monolayer micrographite on TiC(1 1 1)-faceted and TiC(4 1 0) surfaces. Appl. Surf. Sci. 130–132, 876–882.

Tian, H., et al., 2012. Single-layer graphene sound-emitting devices: experiments and modelling. Nanoscale 4, 2272–2277.

Torrisi, F., 2012. Inkjet-printed graphene electronics. ACS Nano 6, 2992–3006.

Tozzini, V., et al., 2013. Prospects for hydrogen storage in graphene. Phys. Chem. Chem. Phys. 15, 80–89.

Traversi, F., et al., 2009. Integrated complementary graphene inverter. Appl. Phys. Lett. 94 (22), 1–3, 223312.

Vaari, J., et al., 1997. The adsorption and decomposition of acetylene on clean and K-covered Co (0001). Catal. Lett. 44 (1–2), 43–49.

Wang, X., et al., 2008. Transparent, conductive graphene electrodes for dye-sensitized solar cells. Nano Lett. 8, 323–327.

Wang, H., et al., 2009. Graphene frequency multipliers. IEEE Electron Device Lett. 30 (5), 547–549.

Wang, J., 2010. Flexible free-standing graphene–silicon composite film for lithium-ion batteries. Electrochem. Commun. 12 (11), 1467–1470.

Wang, X., et al., 2009. N-doping of graphene through electrothermal reactions with ammonia. Science 324 (5928), 768–771.

Wang, Y., et al., 2009. Supercapacitor devices based on graphene materials. J. Phys. Chem. C 113 (30), 13103–13107.

Wang, Y., et al., 2011. Interface engineering of layer-by-layer stacked graphene anodes for high-performance organic solar cells. Adv. Mater. 23, 1514–1518.

Wilmart, Q., et al., 2014. A Klein-tunneling transistor with ballistic graphene. 2D Mater. 1, 1–10, 011006.

Wright, A.R., et al., 2009. Enhanced optical conductivity of bilayer graphene nanoribbons in the terahertz regime. Phys. Rev. Lett. 103, 1–4, 207401.

Wu, J., et al., 2010. Organic light-emitting diodes on solution-processed graphene transparent electrodes. ACS Nano 4, 43–48.

Wu, Q., et al., 2010. Supercapacitors based on flexible graphene/polyaniline nanofiber composite films. ACS Nano 4 (4)), 1963–1970.

Wu, J., et al., 2011. Studies on the electrochemical reduction of oxygen catalyzed by reduced graphene sheets in neutral media. J. Power Sources 196, 1141–1144.

Wu, S., et al., 2011. Application of graphene for preconcentration and highly sensitive stripping voltammetric analysis of organophosphate pesticide. Anal. Chim. Acta 699, 170–176.

Wu, M.S., et al., 2013. Electrophoresis of randomly and vertically embedded graphene nanosheets in activated carbon film as a counter electrode for dye-sensitized solar cells. Phys. Chem. Chem. Phys. 15, 1782–1787.

Wu, S., et al., 2013. Electrochemically reduced graphene oxide and Nafion nanocomposite for ultralow potential detection of organophosphate pesticide. Sens. Actuators B 177, 724–729.

Wu, Y., et al., 2013. Graphene/liquid crystal based terahertz phase shifters. Opt. Express 21, 1–8, 021395.

Wu, B., et al., 2014. Experimental demonstration of a transparent graphene millimetre wave absorber with 28% fractional bandwidth at 140 GHz. Sci. Rep. 4, 1–7, 4130.

Xiao, L., et al., 2012. Crumpled graphene particles for microbial fuel cell electrodes. J. Power Sources 208, 187–192.

Xin, Y., et al., 2011. Preparation and characterization of Pt supported on graphene with enhanced electrocatalytic activity in fuel cell. J. Power Sources 196 (3), 1012–1018.

Yang, X., et al., 2010. Triple-mode single-transistor graphene amplifier and its applications. ACS Nano 4, 5532–5538.

Yin, Z., et al., 2010. Organic photovoltaic devices using highly flexible reduced graphene oxide films as transparent electrodes. ACS Nano 4, 5263–5268.

Yoo, E., et al., 2011. Subnano pt cluster supported on graphene nanosheets for CO tolerant catalysts in polymer electrolyte fuel cells. J. Power Sources 196 (1), 110–115.

Yu, A., et al., 2010. Ultrathin, transparent, and flexible graphene films for supercapacitor application. Appl. Phys. Lett. 96 (25), 1–3, 253105.

Yu, D., et al., 2010. Self-assembled graphene/carbon nanotube hybrid films for supercapacitors. J. Phys. Chem. Lett. 1 (2), 467–470.

Yue, G., et al., 2012. A catalytic composite film of MoS$_2$/graphene flake as a counter electrode for Pt-free dye-sensitized solar cells. Electrochim. Acta 85, 162–168.

Zhang, W.M., et al., 2008. Carbon coated Fe$_3$O$_4$ nanospindles as a superior anode material for lithium-ion batteries. Adv. Funct. Mater. 18, 3941–3946.

Zhang, Y., et al., 2009. Direct observation of a widely tunable bandgap in bilayer graphene. Nature 459, 820–823.

Zhang, Y., 2010. Comparison of graphene growth on single-crystalline and polycrystalline Ni by Chemical Vapor Deposition. J. Phys. Chem. Lett. 1, 3101–3107.

Zhang, Y., et al., 2011. A graphene modified anode to improve the performance of microbial fuel cells. J. Power Sources 196, 5402–5407.

Zhang, Q., et al., 2012. One-step synthesis of graphene/polyallylamine–Au nanocomposites and their electrocatalysis toward oxygen reduction. Talanta 89, 391–395.

Zhou, M., et al., 2009. Electrochemical sensing and biosensing platform based on chemically reduced graphene oxide. Anal. Chem. 81, 5603–5613.

Zhu, L., et al., 2012. DNA electrochemical biosensor based on thionine-graphene nano-composite. Biosens. Bioelectron. 35, 507–511.

CHAPTER 2

Application and Uses of Graphene Oxide and Reduced Graphene Oxide

Sekhar C. Ray

Department of Physics, College of Science, Engineering and Technology, University of South Africa, Florida Park, Johannesburg, South Africa

2.1 INTRODUCTION

Graphite oxide has a layered structure similar to that of graphite, but the plane of carbon atoms in graphite oxide is heavily decorated by oxygen-containing groups, which not only expand the interlayer distance but also make the atomic-thick layers hydrophilic. These oxidized layers could exfoliate in water under ultrasonication. If the exfoliated sheets contain only one or a few layers of carbon atoms like graphene, then these sheets are named graphene oxide (GO) (Novoselov et al., 2004). So, GO is a single-atomic-layered material comprising carbon, hydrogen, and oxygen molecules by the oxidation of graphite crystals, as shown in Figure 2.1 (Stergiou et al., 2014), which are inexpensive and abundant. It is dispersible in water and easy to process. Most importantly, the GO can be (partly) reduced to graphene-like sheets by removing the oxygen-containing groups and with the recovery of a conjugated structure. The reduced GO (rGO) sheets are usually considered one kind of chemically derived graphene and are known as rGO. Some other names have also been given to rGO, such as functionalized graphene, chemically modified graphene, chemically converted graphene, or reduced graphene (Eda et al., 2010). GO has two important characteristics: (i) it can be produced using inexpensive graphite as the raw material and by using cost-effective chemical methods with a high yield and (ii) it is highly hydrophilic and can form stable aqueous colloids to facilitate the assembly of macroscopic structures by simple and cheap solution processes. The graphene sheet consists of only trigonally bonded sp^2 carbon atoms and is perfectly flat (Lui et al., 2009), apart from its microscopic ripples. The heavily decorated GO sheets consist partly of tetrahedrally bonded sp^3 carbon atoms, which are displaced

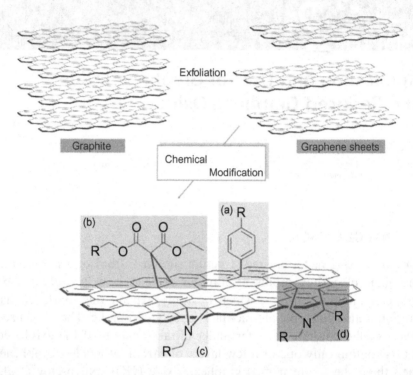

Figure 2.1 General chemical modification routes for exfoliated graphene sheets. (a) [3 + 2] 1,3-dipolar cycloaddition of in situ-generated azomethine ylides, (b) [1 + 2] Bingel cycloaddition, (c) aryl diazonium addition, and (d) azide addition. Reprinted with permission from Stergiou et al., 2014.

slightly above or below the graphene plane (Schniepp et al., 2006). Because of the structure deformation and the presence of covalently bonded functional groups, GO sheets are atomically rough (Paredes et al., 2009; Mkhoyan et al., 2009). Several researchers (Paredes et al., 2009; Kudin et al., 2007; Gomez-Navarro et al., 2007, 2010) have studied the surface of GO and observed highly defective regions, probably due to the presence of oxygen, and other areas are nearly intact. A report shows that the graphene-like honeycomb lattice in GO is preserved, albeit with disorder (i.e., the carbon atoms attached to functional groups are slightly displaced), but the overall size of the unit cell in GO remains similar to that of graphene (Pandey et al., 2008). Hence, GO can be described as a random distribution of oxidized areas with oxygen-containing functional groups combined with nonoxidized regions where most of the carbon atoms preserve sp^2 hybridization. GO and rGO are hot topics in the research and development of graphene, especially regarding mass applications of graphene.

2.2 PREPARATION/SYNTHESIS OF GO/rGO

Graphite is a three-dimensional (3D) carbon-based material comprising millions of graphene layers, whereas graphite oxide is a little different. By oxidation of graphite with strong oxidizing agents, oxygenated functionalities are introduced in the graphite structure, which not only expand the layer separation but also make the material hydrophilic (meaning that they can be dispersed in water).

This property enables the graphite oxide to be exfoliated in water using sonication, ultimately producing single-layer graphene or graphene with a few layers, known as GO. Many modern procedures for the synthesis of GO are based on the method first reported by Hummers in which graphite is oxidized by a solution of potassium permanganate in sulfuric acid (Hummers et al., 1958; Kim et al., 2010). Hydrazine is generally used for the reduction of GO (Gilje et al., 2007). However, hydrazine is highly toxic and can potentially functionalize GO with nitrogen heteroatoms (Shin et al., 2009); because of these issues, alternatives to hydrazine including $NaBH_4$ (Lightcap et al., 2013), ascorbic acid (Fernández-Merino et al., 2010), and HI (Moon et al., 2010; Pei et al., 2010), among others, have been used for the reduction of GO. GO can be reduced to a thin film or in an aqueous solution. GO is effectively a by-product of this oxidization because when the oxidizing agents react with graphite, the interplanar spacing between the layers of graphite is increased. The completely oxidized compound can then be dispersed in a base solution such as water, and GO is then produced. Graphite oxide and GO are very similar chemically, but structurally they are very different. The main difference between graphite oxide and GO is the interplanar spacing between the individual atomic layers of the compounds, which is caused by water intercalation. This increased spacing, caused by the oxidization process, also disrupts the sp^2 bonding network, meaning that both graphite oxide and GO are often described as electrical insulators. GO is a poor conductor but its treatment with light, heat, or chemical reduction can restore most properties of the famed pristine graphene.

To turn graphite oxide into GO, a few methods are possible. The most common techniques are by using sonication, stirring, or a combination of the two. Sonication can be a very time-efficient way of exfoliating graphite oxide, and it is extremely successful at exfoliating graphene; however, it can also heavily damage the graphene flakes, reducing them in surface size from microns to nanometers, and it also produces a wide variety of graphene platelet sizes. The main difference between graphite oxide and GO is the number of layers. Although graphite oxide is a multilayer system in GO dispersion, a few layers of flakes and a monolayer of flakes can be found.

Reducing GO to produce rGO is an extremely vital process because it has a large impact on the quality of the rGO produced; therefore, it will determine how close rGO will come in terms of structure to pristine graphene (Chuang et al., 2014). In large-scale operations where scientific engineers need to utilize large quantities of graphene for industrial applications such as energy storage, rGO is the most obvious solution because of the relative ease in creating sufficient quantities of graphene with desired quality levels. There are a number of ways reduction can be achieved, although they are all methods based on chemical, thermal, or electrochemical means. Some of these techniques are able to produce very high-quality rGO, similar to pristine graphene, but they can be complex or time-consuming to perform.

In the past, scientists have created rGO from GO by:

- Treating GO with hydrazine hydrate and maintaining the solution at 100 for 24 h
- Exposing GO to hydrogen plasma for a few seconds
- Exposing GO to another form of strong pulse light, such as that produced by xenon flashtubes
- Heating GO in distilled water at varying degrees for different lengths of time
- Combining GO with an expansion—reduction agent such as urea and then heating the solution to cause the urea to release reducing gases, followed by cooling
- Directly heating GO to very high levels in a furnace
- Linear sweep voltammetry

Reducing GO by using chemical reduction is a very scalable method; unfortunately, the rGO produced has often resulted in relatively poor yields in terms of surface area and electronic conductibility. Thermally reducing GO at temperatures of $1000°C$ or more creates rGO that has been shown to have a very high surface area, close to that of pristine graphene. The heating process damages the structure of the graphene platelets as pressure builds and carbon dioxide is released.

This also causes a substantial reduction in the mass of the GO, creating imperfections and vacancies, and potentially also has an effect on the mechanical strength of the rGO produced. Electrochemical reduction of GO is a method that has been shown to produce very high-quality rGO, almost identical in terms of structure to pristine graphene. This process involves coating various substrates such as indium tin oxide (ITO) or glass with a very thin layer of GO. Then, electrodes are placed at each end of the substrate, creating a circuit through the GO. In recent experiments, the resulting electrochemically rGO showed a very high carbon to oxygen ratio and also electronic conductivity readings higher than that of silver (8500 S/m compared with ~6300 S/m for silver). Other primary benefits of this technique are that there are no hazardous chemicals used, meaning no toxic waste. Unfortunately, the scalability of this technique has come into question because of the difficulty in depositing GO onto the electrodes in bulk form.

2.3 SURFACE FUNCTIONALIZATION OF GO AND rGO

Once rGO has been produced, there are ways that one can functionalize rGO for use in different applications. By treating rGO with other chemicals or by creating new compounds by combining rGO with other 2D materials, one can enhance the properties of the compound to suit commercial applications. The functionalization of GO not only plays an important role in controlling exfoliation behavior of GO and rGO but also holds the key to various applications.

Covalent functionalization and noncovalent functionalization are two approaches that are used. In covalent functionalization, oxygen functional groups on GO surfaces, including carboxylic acid groups at the edge and epoxy/hydroxyl groups on the basal plane, can be utilized to change the surface functionality of GO. GO has been treated with organic isocyanates to give a number of chemically modified GO. Treatment of isocyanates reduced the hydrophilicity of GO by forming amide and carbamate esters from the carboxyl and hydroxyl groups of GO, respectively. Consequently, isocyanate-modified GO readily formed stable dispersion in polar aprotic solvents, giving completely exfoliated single graphene sheets with a thickness of ~1 nm. This dispersion also facilitated the intimate mixing of the GO sheets with matrix polymers, providing a novel synthesis route to make graphene—polymer nanocomposites. Moreover, modified GO in the suspension could be

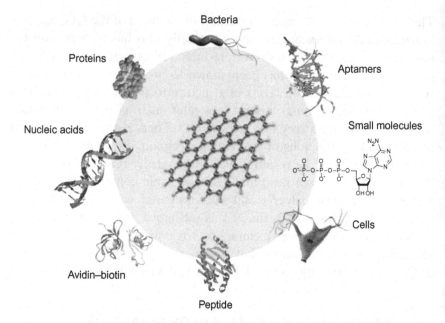

Figure 2.2 Graphene and its derivatives have been reported to be functionalized with avidin–biotin, peptides, NAs, proteins, aptamers, small molecules, bacteria, and cells through physical adsorption or chemical conjugation. Functionalized graphene biosystems with unique properties have been used to build biological platforms, biosensors, and biodevices. Copyright 2015 Elsevier.

chemically reduced in the presence of the host polymer to render electrical conductivity in the nanocomposites (Stankovich et al., 2006; Singh et al., 2011). Figure 2.2 (Wang et al., 2011) shows that graphene and its derivatives have been reported to be functionalized with avidin–biotin, peptides, NAs, proteins, aptamers, small molecules, bacteria, and cells through physical adsorption or chemical conjugation. Functionalized graphene biosystems with unique properties have been used to build biological platforms, biosensors, and biodevices.

2.4 PROPERTIES OF GO AND rGO

One of the advantages of GO is its easy dispersability in water and other organic solvents, as well as in different matrixes, because of the presence of oxygen functionalities. This remains a very important property when mixing the material with ceramic or polymer matrixes when trying to improve their electrical and mechanical properties. However, in terms of electrical conductivity, GO is often described as an electrical insulator because of the disruption of its sp^2 bonding networks.

To recover the honeycomb hexagonal lattice, and with it the electrical conductivity, the reduction of GO has to be achieved. It has to be taken into account that once most of the oxygen groups are removed, the rGO obtained is more difficult to disperse because of its tendency to create aggregates. Functionalization of GO can fundamentally change the properties of GO. The resulting chemically modified graphenes could then potentially become much more adaptable for many applications. There are many ways in which GO can be functionalized, depending on the desired application. For optoelectronics, biodevices, or as a drug-delivery material, it is possible to substitute amines for the organic covalent functionalization of graphene to increase the dispersability of chemically modified graphenes in organic solvents. It has also been proven that porphyrin-functionalized primary amines and fullerene-functionalized secondary amines could be attached to GO platelets, ultimately increasing nonlinear optical performance. For GO to be usable as an intermediary in the creation of monolayer or few-layer graphene sheets, it is important to develop an oxidization and reduction process that is able to separate individual carbon layers and then isolate them without modifying their structure. So far, although the chemical reduction of GO is currently seen as the most suitable method of mass production of graphene, it has been difficult for scientists to complete the task of producing graphene sheets of the same quality as mechanical exfoliation but on a much larger scale. Once this issue is overcome, we can expect to see graphene become much more widely used in commercial and industrial applications.

2.5 APPLICATIONS OF GO AND rGO

2.5.1 GO/rGO in Electronics Devices

Several electronic devices have been fabricated using GO as a starting material for at least one of the components. One such device is a graphene-based field effect transistor (FET) (Su, 2010; Wang, S. et al., 2010). FETs that use rGO have been used as chemical sensors (Lu et al., 2011; Chen et al., 2012; He et al., 2012) and biosensors. FETs that use functionalized rGO as the semi-conductor have been used as biosensors to detect hormonal catecholamine molecules (He et al., 2010), avidin (He et al., 2011), and DNA (Cai et al., 2014). Liu et al. (2010) studied the electrochemical glucose sensor using GO functionalized with glucose oxidase after being deposited on an electrode. One of the major areas where GO can be expected to be used is in the production of transparent conductive films after being deposited on any substrate. Such coatings could be used in flexible electronics, solar cells, liquid crystal devices, chemical sensors, and touch screen devices. Cai et al. (2014), Matyba et al. (2010), and Becerril et al. (2008) used GO/rGO as a transparent electrode for light-emitting diodes (LEDs) and solar cell devices. The transparent electrode GO/rGO has also been used as a hole transport layer in polymer solar cells and LEDs (Saha et al., 2014; Li et al., 2010). Different electronic devices are shown in Figure 2.3.

2.5.2 GO/rGO as Energy Storage Device

GO and rGO have an extremely high surface area; therefore, these materials are considered for usage as electrode materials in batteries and double-layered capacitors, as well as fuel cells and solar cells (Zhu et al., 2010a,b). Production of GO can be easily scaled-up compared with other graphene materials, and therefore it may soon be used for energy-related purposes. Its ability to store hydrogen may, in the future, prove very useful for the storage of hydrogen fuel in hybrid cars. Nanocomposites of GO/rGO can also be used for high-capacity energy storage in lithium ion batteries. In this case, electrically insulating metal oxide nanoparticles were adsorbed onto rGO to increase the performance of these materials in batteries (Wang, H. et al., 2010; Yang et al., 2010; Lee et al., 2010; Zhou et al., 2010; Zhang et al., 2010). Zhou et al. (2010) fabricated the Li–ion battery device using rGO-wrapped Fe_3O_4 anode material (i.e., Fe_3O_4 on rGO) and found that the energy storage capacity and cycle stability are increased compared with pure Fe_3O_4 or Fe_2O_3. Zhu et al. (2010a,b, 2011) made high-surface-area rGO using microwave-assisted exfoliation and, hence, reducing GO for the fabrication of super-capacitors as energy storage device. Bo et al. (2014) fabricated electronic gas sensors and super-capacitors with the caffeic acid (CA)-rGO and found good performance for potential sensing and energy storage applications. Different bioapplications are shown in Figure 2.4.

2.5.3 GO/rGO as Biosensors

GO/rGO is a fluorescent material that could be used for biosensing applications, for early disease detection, and even for assisting in finding cures for cancer and detecting biologically relevant molecules. GO has been successfully used in fluorescent-based biosensors for the detection of DNA and proteins with a promise of better diagnostics

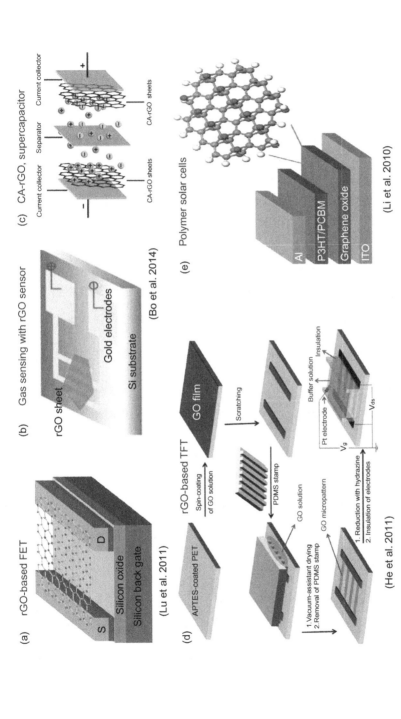

(a) rGO-based FET

(b) Gas sensing with rGO sensor

(c) CA-rGO, supercapacitor

(d) rGO-based TFT

(e) Polymer solar cells

Figure 2.3 (a) rGO-based FET. (b and c) gas sensing with rGO and CA-rGO-based super-capacitor, (d) rGO-based TFT, and (e) polymer solar cells. Reprinted with permission from Lu et al., 2011; Bo et al., 2014; He et al., 2011; and Li et al., 2010. Copyright 2015 American Chemical Society (a), (d), (e) and Nature publishing group (b), (c).

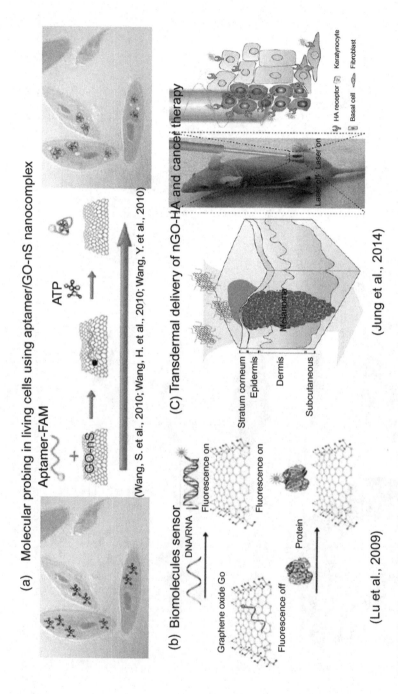

(a) Molecular probing in living cells using aptamer/GO-nS nanocomplex

Aptamer-FAM

+ GO-nS

ATP

(Wang, S. et al., 2010; Wang, H. et al., 2010; Wang, Y. et al., 2010)

(b) Biomolecules sensor

DNA/RNA

Graphene oxide Go

Fluorescence on

Fluorescence off

Fluorescence on

Protein

(Lu et al., 2009)

(C) Transdermal delivery of nGO-HA and cancer therapy

Laser off   Laser on

HA receptor

Keratinocyte

Basal cell

Fibroblast

Stratum corneum

Epidermis

Dermis

Subcutaneous

Melanoma

(Jung et al., 2014)

Figure 2.4 *(a) Molecular probing in living cells, (b) biomolecules sensor, and (c) transdermal delivery of nGO-HA and cancer therapy.* Reprinted with permission from Wang, S. et al., 2010; Wang, H. et al., 2010; Wang, Y. et al., 2010; Lu et al., 2009; and Jung et al., 2014. Copyright 2015 (a), (c) American Chemical Society and (b) Wiley Publishing Company.

for HIV. GO has been used as a fluorescence quenching material in biosensors that utilize the fluorescence resonance energy transfer (FRET) effect. Wang, Y. et al. (2010) used the FRET effect in a fluorescein-labeled ATP aptamer to sense ATP as low as 10 μM. Lu et al. (2009) used single-stranded DNA (ssDNA) with a fluorescence tag and found that it bound noncovalently to GO with subsequent quenching of the fluorescence of the tag. Addition of a complementary ssDNA removed the tagged DNA from the GO surface and restored the fluorescence. Song et al. (2011) used folic acid-functionalized GO to detect human cervical cancer and human breast cancer cells.

2.5.4 GO/rGO as Biomedical Applications

GO is used in the biomedical field, particularly in drug-delivery systems. GO is likely superior to many other anticancer drugs because it does not target healthy cells, only tumors, and has a low toxicity (Yang et al., 2011). Functionalized nano-GO (nGO) has been used in several studies on targeted delivery of anticancer drugs. Polyethylene glycol (PEG)-functionalized nGO with SN38, a campothecin derivative adsorbed onto the surface (nGO−PEG−SN38) that was used as a water-soluble and serum-soluble source of the drug (Liu et al., 2008). In this study, nGO−PEG−SN38 was shown to be three orders of magnitude more effective than irinotecan (CPT-11), an FDA-approved SN38 prodrug, at reducing the cell viability of human colon cancer cell line HTC-116 (Liu et al., 2008). The effectiveness of nGO−PEG−SN38 was similar to SN38 in DMSO (Liu et al., 2008). Melanoma skin cancer in mice has been treated using photothermal ablation therapy with a near-infrared laser and nGO that was functionalized with PEG and hyaluronic acid and delivered transdermally (Jung et al., 2014). In another study, magnetite was adsorbed onto GO loaded with the anticancer drug doxorubicin hydrochloride for targeted delivery of the drug to specific sites using magnets (Yang et al., 2009). Shen et al. (2012) studied various biomedical applications using GO/rGO, particularly in drug delivery, cancer therapy, and biological imaging.

2.5.5 GO as Water Purification (Filter)

Permeation of water through the membrane was attributed to swelling of GO structures, which enables a water penetration path between individual GO layers. The interlayer distance of dried Hummers graphite oxide was reported as 6.35 Å, but in liquid water it increased to 11.6 Å. The permeation rate of the membranes for the water is

0.1 mg/min/cm^2, and the diffusion rate of water is 1 cm/h. These oxides also could be used as cation exchange membrane KCl, HCl, CaCl$_2$, MgCl$_2$, and BaCl$_2$ solutions. The membranes were also reported to be permeable by large alkaloid ions because they are able to penetrate between GO layers (Boehm et al., 1960). GO membranes were also actively studied in the 1960s for application in water desalination, but they were never used for practical applications (Joshi, 2014). Retention rates more than 90% were reported in this study for NaCl solutions using stabilized GO membranes in a reverse osmosis setup. GO membranes could be used for the filtration of sea water. GO film is super thin (just one single atom thick), so the water simply "pops through the very, very small holes that are in the graphene and leaves the salt behind." GO film is 500-times thinner than the best filter on the market and ~1000-times stronger than steel, but its permeability is ~100-times greater than the best competitive membrane on the market. The specimens allow ions from common salts to pass through the filter but retain some larger ions (Joshi, 2014). Narrow capillaries allow rapid permeation by monolayer or bilayer water. Helium cannot pass through the membranes in humidity-free conditions, but it penetrates easily when exposed to humid gas, whereas water vapor passes without resistance. Dry laminates are vacuum-tight, but when immersed in water they act as molecular sieves, blocking some solutes with hydrated radii larger than 4.5 Å.

2.5.6 GO/rGO as Coating Technology

Multilayer GO films are optically transparent and impermeable under dry conditions. Exposed to water (or water vapor), they allow passage of molecules smaller than a certain size. The films consist of millions of randomly stacked flakes, leaving nano-sized capillaries between them. Closing these nanocapillaries using chemical reduction with hydro-iodic acid creates rGO films that are completely impermeable to gases, liquids, or strong chemicals >100 nm thick. Glassware or copper plates covered with such a graphene "paint" can be used as containers for corrosive acids. Graphene-coated plastic films could be used in medical packaging to improve shelf life.

2.5.7 GO/rGO Composites and Paper-Like Materials

GO mixes readily with many polymers, forming nanocomposites, and greatly enhances the properties of the original polymer, including elastic modulus, tensile strength, electrical conductivity, and thermal stability. In its solid form, GO flakes tend to attach to one another, forming thin and extremely stable paper-like structures that can be folded, wrinkled, and stretched. Such free-standing GO films are considered for applications including hydrogen storage applications, ion conductors, and nanofiltration membranes.

2.6 CONCLUSION AND PERSPECTIVES OF GO/rGO

As in many other fields, research of GO/rGO applications has seen dramatic progress and it is expanding rapidly. The advances made in this area so far are exciting and encouraging; however, the challenges are also huge and must be overcome. One such challenge is thorough: a profound understanding of graphene–cell (or tissue or organ) interactions, especially the cellular uptake mechanism. Such knowledge certainly would facilitate the development of more efficient GO-based nanoplatforms for drug delivery, biosensing, and other applications. The toxicity of graphene and GO, at in vitro and in vivo levels, is another major concern. The mechanisms of the in vitro biotoxicity caused by graphene are related to oxidative stress and damage of cell membranes. Clearly, a systematic study is highly desired to address the safety concerns before the practical application of graphene in biomedicine. These goals can only be reached by joint efforts from chemistry, biomedicine, materials sciences, and nanotechnology. Development of suitable chemical synthesis and functionalization approaches for precise control over size, size distribution, morphology, structural defects, and oxygen-containing groups of GO is urgently needed, because these are closely correlated to the performance of the GO-based nanomaterials for biomedical applications and safety issues. For GO-based biosensing based on the FRET principle, tuning electronic property of GO by controllable modification and reduction of originally prepared GO, and development of techniques to integrate GO into practical devices with high sensitivity, selectivity with acceptable reproducibility, reliability, and low cost remain big challenges. As for GO-based bioimaging, although the GQDs have distinct advantages over II–IV QDs in their intrinsic biocompatibility, safety, and easy functionalization, the weak fluorescence intensity (with quantum yield ~10%) and broad emission band (with bandwidth beyond 100 nm) are certainly major obstacles for their use in biodetection and labeling; therefore, more efforts must be made to prepare GQDs with good control of size and size distribution, surface defects, and functionalization to improve fluorescence quantum yield and other important properties. The research on graphene/GO-based scaffold materials for cell culture is relatively new and deserves special attention. Studies in this field so far have demonstrated that graphene and GO are able to accelerate the growth, differentiation, and proliferation of stem cells, and therefore hold great promise in tissue engineering, regenerative medicine, and other biomedical fields.

2.7 THE PRESENT CHALLENGES AND FUTURE RESEARCH IN GO/rGO NANOMATERIAL

We have reviewed the different applications of GO/rGO. Although the full reduction of GO to graphene is still difficult to achieve, partial reduction of GO is rather easy. The structure and chemistry of GO/rGO are discussed briefly, which may be helpful in promoting the uses as well as the scientific understanding of the nature of GO/rGO. However, GO sheets with a high concentration of lattice defects are difficult to fully deoxygenate, and the defects themselves are difficult to heal by posttreatment. As a result, controllable oxidation during the production of GO is needed to achieve highly reducible GO. Future research on GO/rGO should mainly focus on two topics: (i) a much deeper understanding of the reduction mechanism and (ii) how to control the oxidation of graphite and the reduction of GO. This is because controllable functionalization that can alter the properties of graphene to fulfill specific requirements in applications is equally important to obtain nondefective graphene and, for example, to change the gapless semi-metallic graphene into a semi-conductor with a proper band gap. The previous research on GO and rGO has inspired a possible way to achieve such changes so that GO and rGO show obvious semi-conductor-like properties (Eda et al., 2010). Research on the oxidation and reduction combined with a deep understanding of graphene structure may allow us to realize good control of the attaching and elimination of functional groups to some specific locations on the carbon plane. Further research on the controllable oxidation and reduction of graphene may facilitate the applications of graphene as semi-conductors used in transistor and photoelectronic devices.

REFERENCES

Becerril, H.A., Mao, J., Liu, Z., Stoltenberg, R.M., Liu, Z., Chen, Y., et al., 2008. Evaluation of solution-processed reduced graphene oxide films as transparent conductors. ACS Nano 2008 (2), 463–470.

Bo, Z., et al., 2014. Green preparation of reduced graphene oxide for sensing and energy storage applications. Sci. Rep. 4 (4684), 1–8.

Boehm, H.P., et al., 1960. Graphite oxide and its membrane properties. J. Chim. Phys. Rev. Gen. Colloides 58 (12), 110–117.

Cai, B., et al., 2014. Ultrasensitive label-free detection of PNA–DNA hybridization by reduced graphene oxide field-effect transistor biosensor. ACS Nano 8, 2632–2638.

Chen, K., et al., 2012. Hg(II) ion detection using thermally reduced graphene oxide decorated with functionalized gold nanoparticles. Anal. Chem. 84, 4057–4062.

Chuang, C.-H., et al., 2014. The effect of thermal reduction on the photoluminescence and electronic structures of graphene oxides. Sci. Rep. 4 (4525), 1–7.

Eda, G., et al., 2010. Chemically derived graphene oxide: towards large-area thin-film electronics and optoelectronics. Adv. Mater. 22 (22), 2392–2415.

Fernández-Merino, M.J., et al., 2010. Vitamin C is an ideal substitute for hydrazine in the reduction of raphene oxide suspensions. J. Phys. Chem. C 114, 6426–6432.

Gilje, S., et al., 2007. A chemical route to graphene for device applications. Nano Lett. 7, 3394–3398.

Gomez-Navarro, C., et al., 2007. Electronic transport properties of individual chemically reduced graphene oxide sheets. Nano Lett. 7 (11), 3499–3503.

Gomez-Navarro, C., et al., 2010. Atomic structure of reduced graphene oxide. Nano Lett. 10 (4), 1144–1148.

He, Q., et al., 2010. Centimeter-long and large-scale micropatterns of reduced graphene oxide films: fabrication and sensing applications. ACS Nano 4, 3201–3208.

He, Q., et al., 2011. Transparent, flexible, all-reduced graphene oxide thin film transistors. ACS Nano 5, 5038–5044.

He, Q., et al., 2012. Graphene-based electronic sensors. Chem. Sci. 2012 (3), 1764–1772.

Hummers, W.S., et al., 1958. Preparation of graphitic oxide. J. Am. Chem. Soc. 80, 1339.

Joshi, R.K., 2014. Precise and ultrafast molecular sieving through graphene oxide membranes. Science 343 (6172), 752–754.

Jung, H.S., 2014. Nanographene oxide hyaluronic acid conjugate for photothermal ablation therapy of skin cancer. ACS Nano 8, 260–268.

Kim, J., et al., 2010. Graphene oxide sheets at interfaces. J. Am. Chem. Soc. 132, 8180–8186.

Kudin, K.N., et al., 2007. Raman spectra of graphite oxide and functionalized graphene sheets. Nano Lett. 8 (1), 36–41.

Lee, J.K., et al., 2010. Silicon nanoparticles–graphene paper composites for Li ion battery anodes. Chem. Commun. 46, 2025–2027.

Li, S-S., et al., 2010. Solution-processable graphene oxide as an efficient hole transport layer in polymer solar cells. ACS Nano 4, 3169–3174.

Lightcap, I., et al., 2013. Graphitic design: prospects of graphene-based nanocomposites for solar energy conversion, storage, and sensing. Acc. Chem. Res. 2013 (46), 2235–2245.

Liu, Z., et al., 2008. PEGylated nano-graphene oxide for delivery of water-insoluble cancer drugs. J. Am. Chem. Soc. 130, 10876–10877.

Liu, Y., et al., 2010. Biocompatible graphene oxide-based glucose biosensors. Langmuir 26, 6158–6160.

Lu, C.-H., et al., 2009. A graphene platform for sensing biomolecules. Angew. Chem. Int. Ed. 48, 4785–4786.

Lu, G., et al., 2011. Toward practical gas sensing with highly reduced graphene oxide: a new signal processing method to circumvent run-to-run and device-to-device variations. ACS Nano 5, 1154–1164.

Lui, C.H., et al., 2009. Ultra flat graphene. Nature 462 (7271), 339–341.

Matyba, P., et al., 2010. Graphene and mobile ions: the key to all-plastic, solution-processed light-emitting devices. ACS Nano 4, 637642.

Mkhoyan, K.A., et al., 2009. Atomic and electronic structure of graphene-oxide. Nano Lett. 9 (3), 1058–1063.

Moon, K., et al., 2010. Reduced graphene oxide by chemical graphitization. Nat. Commun. 1, 73–78.

Novoselov, K.S., et al., 2004. Electric field effect in atomically thin carbon films. Science 306 (5696), 666–669.

Pandey, D., et al., 2008. Scanning probe microscopy study of exfoliated oxidized graphene sheets. Surf. Sci. 602 (9), 1607–1613.

Paredes, J.I., et al., 2009. Atomic force and scanning tunnelling microscopy imaging of graphene nanosheets derived from graphite oxide. Langmuir 25 (10), 5957–5968.

Pei, S., et al., 2010. Direct reduction of graphene oxide films into highly conductive and flexible graphene films by hydrohalic acids. Carbon 48, 4466–4474.

Saha, S.K., et al., 2014. Solution-processed reduced graphene oxide in light-emitting diodes and photovoltaic devices with the same pair of active materials. RSC Adv. 2014 (4), 35493–35499.

Schniepp, H.C., et al., 2006. Functionalized single graphene sheets derived from splitting graphite oxide. J. Phys. Chem. B 110 (17), 8535–8539.

Shen, H., et al., 2012. Biomedical applications of graphene. Theranostics 2 (3), 283–294.

Shin, H-J., et al., 2009. Efficient reduction of graphite oxide by sodium borohydride and its effect on electrical conductance. Adv. Funct. Mater. 19, 1987–1992.

Singh, V., et al., 2011. Graphene based materials: past, present and future. Prog. Mater. Sci. 56, 1178–1271.

Song, Y., et al., 2011. Selective and quantitative cancer cell detection using target-directed functionalized graphene and its synergetic peroxidase-like activity. Chem. Commun. 47, 4436–4438.

Stankovich, S., et al., 2006. Synthesis and exfoliation of isocyanate-treated graphene oxide nanoplatelets. Carbon 44, 3342–3347.

Stergiou, A., et al., 2014. Donor–acceptor graphene-based hybrid materials facilitating photoinduced electron-transfer reactions. Beilstein J. Nanotechnol. 5, 1580–1589.

Su, C-Y., et al., 2010. Highly efficient restoration of graphitic structure in graphene oxide using alcohol vapors. ACS Nano 4, 5285–5292.

Wang, S., et al., 2010. High mobility, printable, and solution-processed graphene electronics. Nano Lett. 10, 92–98.

Wang, H., et al., 2010. Mn_3O_4-graphene hybrid as a high-capacity anode material for lithium ion batteries. J. Am. Chem. Soc. 132, 13978–13980.

Wang, Y., et al., 2010. Aptamer/graphene oxide nanocomplex for in situ molecular probing in living cells. J. Am. Chem. Soc. 132, 9274–9276.

Wang, Y., et al., 2011. Graphene and graphene oxide: biofunctionalization and applications in biotechnology. Trends Biotechnol. 29 (5), 205−212.

Yang, X., et al., 2009. Superparamagnetic graphene oxide−Fe_3O_4 nanoparticles hybrid for controlled targeted drug carriers. J. Mater. Chem. 2009 (19), 2710−2714.

Yang, S., et al., 2010. Fabrication of graphene-encapsulated oxide nanoparticles: towards high-performance anode materials for lithium storage. Angew. Chem. Int. Ed. 49, 8408−8411.

Yang, X., et al., 2011. Multi-functionalized graphene oxide based anticancer drug-carrier with dual-targeting function and pH-sensitivity. J. Mater. Chem. 21, 3448−3454.

Zhang, M., et al., 2010. Magnetite/graphene composites: microwave irradiation synthesis and enhanced cycling and rate performances for lithium ion batteries. J. Mater. Chem. 20, 5538−5543.

Zhou, G., et al., 2010. Graphene-wrapped Fe_3O_4 anode material with improved reversible capacity and cyclic stability for lithium ion batteries. Chem. Mater. 22, 5306−5313.

Zhu, Y., et al., 2010a. Microwave assisted exfoliation and reduction of graphite oxide for ultracapacitors. Carbon 48, 2118−2122.

Zhu, Y., et al., 2010b. Graphene and graphene oxide: synthesis, properties, and applications. Adv. Mater. 22 (35), 3906−3924.

Zhu, Y., et al., 2011. Carbon-based supercapacitors produced by activation of graphene. Science 332, 1537−1541.

CHAPTER 3

Graphene-Based Carbon Nanoparticles for Bioimaging Applications

Nikhil R. Jana[1] and Sekhar C. Ray[2]

[1]Centre for Advanced Materials, Indian Association for the Cultivation of Science (IACS), Jadavpur, Kolkata, India; [2]Department of Physics, College of Science, Engineering and Technology, University of South Africa, Florida Park, Johannesburg, South Africa

3.1 INTRODUCTION

The fluorescence carbon nanoparticle (FCN) is considered a new-generation green nanomaterial (Baker et al., 2010; Wang, X. et al., 2010; Li, Q. et al., 2010; Cao, L. et al., 2013; Krysmann et al., 2012; Yu et al., 2012; Guo et al., 2012; Zhu et al., 2012) and a promising alternative for fluorescent semiconductor nanocrystals. FCN has been demonstrated as a potential optical detection probe (Zhu et al., 2012), bioimaging probe (Li, Q. et al., 2010), light-emitting diode material (Guo et al., 2012), and efficient visible light-active photocatalyst (Yu et al., 2012), and has been used in other optoelectronic device applications (Gruber et al., 1997; Neugart et al., 2007; Batalov et al., 2009; Glinka et al., 1999; Zyubin et al., 2009; Yu et al., 2005; Sun et al., 2006; Zhou et al., 2007; Fu et al., 2007; Wee et al., 2007; Lim et al., 2009; Cao, L. et al., 2007; Liu et al., 2007; Zhao et al., 2008; Selvi et al., 2008; Bourlinos et al., 2008; Mochalin and Gogotsi, 2009; Liu et al., 2009). These carbon nanoparticles are biocompatible and chemically inert (Sun et al., 2006; Zhou et al., 2007; Fu et al., 2007; Wee et al., 2007; Lim et al., 2009; Ushizawa et al., 2002; Cahalan et al., 2002; Huang et al., 2004; Kong et al., 2005a,b), which gives advantages over conventional cadmium-based quantum dots (Medintz et al., 2005; Kamat, 2008) and/or nanostructure materials. However, FCN is relatively less explored compared with other carbon-based materials such as fullerene (Diederich, 1996), carbon nanotube (Tasis et al., 2006), and grapheme (Dreyer et al., 2010). In addition, the understanding of the origin of fluorescence in the carbon nanoparticle is far from sufficient (Sun et al., 2006; Zhou et al., 2007; Cao, L. et al., 2007; Zhao et al., 2008). Information on the microstructure and surface ligands remains unclear and details of the organic passivation is not sufficient to aid understanding of the surface states beneficial for light emission. FCN has been synthesized via physical methods such as high-energy radiation-based creation of point defect in diamond (Gruber et al., 1997; Neugart et al., 2007; Batalov et al., 2009; Yu et al., 2005) and laser ablation of graphite (Sun et al., 2006; Cao, L. et al., 2007), or chemical methods like oxidation of candle soot (Liu et al., 2007; Ray et al., 2009), carbonization of carbohydrate (Selvi et al., 2008; Peng and Travas-Sejdic, 2009; Zhang, B. et al., 2010; Zhang, J. et al., 2010; Yang et al., 2011, 2012), thermal decomposition of small molecules (Selvi et al., 2008; Bourlinos et al., 2008a,b; Liu et al., 2009), pyrolysis of polymers (Liu et al., 2009), microwave-based pyrolysis (Li et al., 2011; Wang et al., 2011; Chandra et al., 2012), P_2O_5-based room temperature dehydration of small molecule (Fang et al., 2012), electrochemical method (Li, H. et al., 2010; Bao et al., 2011), and chemical breakdown of carbon fiber (Peng et al., 2012), graphene (Zhuo et al., 2012), and graphite (Zhao et al., 2008; Hens et al., 2012). A wide range of fluorescent carbon particles (CPs) of different colors can be prepared using these approaches. However, there are four distinct limitations of currently available FCN. First, most of the synthetic methods produce weakly fluorescent FCN with <1% quantum yield. Second, no methods are currently available for large-scale synthesis of high-quality FCN. Although some methods report milligram-scale FCN with 5–60% quantum yield, they require sophisticated high-energy radiation-based synthesis followed by surface functionalization and size separation (Wang, F. et al., 2010). Third, although many methods report blue-/green-emitting FCN, only a few methods report yellow-emitting and red-emitting FCNs. In addition yellow-/red-emitting FCNs are generally mixed with blue-/green-emitting FCNs and need to be isolated via specialized size separation methods (Wang, F. et al., 2010; Liu et al., 2007; Li, H. et al., 2010; Hens et al., 2012). Fourth, although functionalization of nanoprobes is essential to enhance the labeling specificity, such strategies are little developed for FCNs (Li, H. et al., 2011; Yu et al., 2005; Selvi et al., 2008; Ray et al., 2009; Bourlinos et al., 2008a,b; Fang et al., 2012; Peng et al., 2012). In addition, the synthetic methods are cumbersome and inefficient. Recent reports showed that surface passivation can lead to a significant increase in fluorescence quantum yield (4–15%); however, the exact mechanism is not yet clear (Sun et al., 2006; Liu et al., 2009). So, more efficient and large-scale synthesis of FCN and their isolation, purification, functionalization, and, hence, bioimaging and applications are very challenging. Here, we describe some of the synthetic methods and functionalization approaches that we have developed in the context of the current development of the field and also highlight some important approach of others.

3.2 PREPARATION PROCESS OF CARBON NANOPARTICLES

3.2.1 Synthesis of CP from Oxidation of Burning Candle Soot

Liu et al. (2007) and Ray et al. (2009) used the burning candle soot for the synthesis of FCN for bioimaging and other device applications. Scheme 3.1 shows the different steps in the preparation of FCN from soot. In this process, powder carbon soot (collected from burning candle) was mixed with nitric acid (5 M) and refluxed at 100°C for 12 h. The black solution was cooled and centrifuged at 3000 rpm to separate out unreacted carbon soot. The light brown-yellow supernatant was collected that showed green fluorescence under UV exposure. The aqueous supernatant was mixed with acetone (water:acetone volume ratio was 1:3) and centrifuged at 14,000 rpm for 10 min (Rat et al., 2009). The black precipitate was collected and dissolved in 5−10 mL water. The colorless and nonfluorescent supernatant was discarded. This step of purification separates excess nitric acid from the carbon nanoparticles. This concentrated aqueous solution with almost neutral pH was taken for further use.

3.2.2 Synthesis of Carbon Nanoparticle from Carbohydrate Carbonization Method

The well-known carbohydrate carbonization method usually produces nonfluorescent CPs of >10 nm size (Sun and Li, 2004) or weakly fluorescent blue−green-emitting CPs (Peng and Travas-Sejdic, 2009). However, Bhunia et al. (2013) extensively varied the carbonization condition to

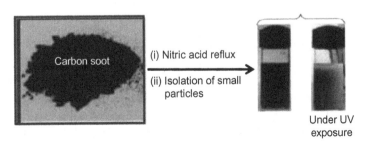

Scheme 3.1 Steps in the preparation of fluorescent carbon nanoparticles (CNPs) from Soot. From Ray et al., 2009. Copyright 2015 the American Chemical Society.

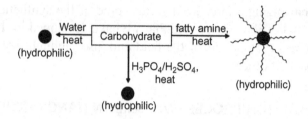

Scheme 3.2 Different controlled carbonization approach of carbohydrate in producing fluorescent carbon nanoparticle. From Bhunia et al., 2013. Copyright 2015 Nature publishing group.

finely tune the nucleation growth so that carbon nanoparticles <10 nm in size can be prepared with narrow size distribution. Bhunia et al. (2013) followed two different approaches that have been used to control the nucleation growth as shown in Scheme 3.2. In the first approach, different carbohydrate molecules have been used as carbon sources that dehydrate in different temperature. In the second approach, the dehydrating agent and reaction temperature are varied to control the nucleation-growth kinetics. This synthetic condition is similar to the well-established synthetic condition for metal/metal oxide/semiconductor nanoparticles (Jana et al., 2004) and produces hydrophobic FCN with fatty amine capping. This approach is most suitable for synthesis of blue-emitting and green-emitting FCNs. The best condition for yellow- and red-emitting FCNs is concentrated phosphoric acid-based carbonization of carbohydrate at 80–90°C. The presented synthetic method restricts particle size <10 nm and provides appropriate chemical composition of FCN for tunable emission with high-fluorescence quantum yield. Those FCNs have tunable emission (such as blue, green, yellow, and red) with quantum yield between 6% and 30% and have been synthesized in milligram to gram scale, and they have transformed into various functional nanoprobes for different biological labeling and imaging applications. Different synthetic methods along with the quantum yield of FCNs are summarized in Table 3.1.

3.2.3 Functionalization of FCN

Water-soluble FCNs with primary amine on their surface have been used by Bhunia et al. for covalent functionalization with affinity molecules. Hydrophobic FCNs that are water-insoluble and do not have such amine functionality have been transformed into water-soluble and primary amine-terminated FCNs using polymer coating. In a typical procedure of polymer coating, 1 mL chloroform solution of hydrophobic FCNs (~10 mg/mL) was mixed with 0.5 mL chloroform solution of

Table 3.1 Different Synthetic Methods Along with Quantum Yield for FCNs				
Method	Emission (QY)	Size (nm)	Limitations	Ref.
High energy radiation of diamond	Red, NIR (−)	>100	Large size, one color, difficult for large-scale synthesis	Gruber et al. (1997), Yu et al. (2005)
Laser ablation of graphite and electric discharge of carbon	Blue to red (up to 60%)	<10	Surface passivation step and size separation is necessary, yellow/red-emitting particles have very weak emission, difficult to make in milligram scale	Wang, F. et al., (2010), Wang, X. et al. (2010), Sun et al. (2006)
Oxidation of candle soot	Blue to red (<3%)	<10	Difficulty of size separation, difficult to make in milligram scale, yellow/red-emitting particles have weak emission	Liu et al. (2007), Ray et al. (2009)
Carbonization of carbohydrate	Blue, green (up to 10%)	<10	Difficult to make yellow/red-emitting particle, mixed with all emission colors	Selvi et al. (2008), Peng and Travas-Sejdic (2009), Zhang, B. et al. (2010), Zhang, J. et al. (2010), Yang et al. (2011), Yang et al. (2012)
Carbonization of small molecules	Blue, green (5−50%)	<10	Yellow/red-emitting particles are difficult to make, mixed with all emission colors	Bourlinos et al. (2008a,b), Wang, F. et al., (2010), Wang, X. et al. (2010)
Carbonization of polymer	Blue to red (15% for blue)	<10	Low quantum yield for yellow/red emission, mixed with all emission colors	Liu et al. (2009)
Electrochemical	Blue to yellow (up to 12%)	1−5	Difficult to make in large-scale, red-emitting particles are difficult to make	Li, H. et al., 2010; Li, Q. et al., 2010, Bao et al. (2011)
Chemical breakdown of carbon fiber/ graphene/graphite	Blue, green (0.5−28%)	<100	Difficult to make yellow/red-emitting particle	Cao et al. (2012), Peng and Travas-Sejdic (2009), Hens et al. (2012)

Source: *From Bhunia et al., 2013. Copyright 2015 by the Nature publishing group.*

polymaleicanhydride-1-octadecene solution (80 mg dissolved in 0.5 mL chloroform) and sonicated for 5 min. Next, chloroform solution of *O,O′*-bis(2-amino propyl) polypropylene glycol-block-polyethylene glycol-block-polypropylene glycol (PEG-diamine) (15 mg PEG-diamine dissolved in 0.5 mL chloroform) was mixed with FCN solution. After being kept overnight, the whole solution was evaporated to dryness and 2−3 mL aqueous Na_2CO_3 solution was added and kept at room temperature for 1−2 h. FCNs were completely dissolved in water, which has been used for functionalization. Functionalization of primary amine-terminated FCN has been achieved by Bhunia et al. using well-known bioconjugation chemistry. Folic acid functionalization of FCN was achieved by using the N-hydroxysuccinimide-folate-based approach. Typically, aqueous solution of FCN was prepared in borate buffer of pH 9 and mixed with dimethylformamide solution of N-hydroxysuccinimide-folate. After overnight reaction, the solution was dialyzed using a dialysis membrane (MWCF 12–14 kDa) to remove excess reagents.

TAT peptide (CHHHHHHHHHHHGRKKRRQRRR, MW 2871, 95% purity) functionalization was achieved using well-known 4-maleimido-butyric acid N-hydroxysuccinimide ester reagent. The maleimido group of this reagent reacts with the thiol group of cysteine-terminated TAT peptide, and N-hydroxysuccinimide group reacts with the primary amine group of FCN. This results in covalent linking between FCN and TAT peptide. Typically, the aqueous solution of FCN was mixed with dimethylformamide solution of conjugation reagent and after 30 min TAT peptide solution was added to it. After overnight reaction, the solution was dialyzed using dialysis membrane (MWCF 12–14 kDa) to remove excess reagents and unbound peptide.

3.3 PROPERTIES OF CARBON NANOPARTICLES

3.3.1 Physical and Structural Properties

Natural available carbon is black and completely insoluble in water even after ultrasonication. This is because they are large in size and hydrophobic in nature. When this carbon (for example, candle soot) is refluxed with nitric acid, light brown supernatant solution is obtained along with an insoluble black precipitate. Brown-yellow supernatant indicates a part of CP becomes small and water-soluble during the refluxing processes. This soluble particle exhibits green fluorescence when irradiated with UV light, whereas the precipitate part shows no fluorescence. The nature of fluorescence spectra strictly depends on different sizes/types of particles with different colors. So, identification/separation of FCNs from the mixture of particles is very important. Liu et al. (2007) separated these particles by the gel electrophoresis technique. Ray et al. (2009) performed the size separation of particles to isolate FCNs of different fluorescent properties. The size separation has been performed in a solvent mixture with high-speed centrifuge-based separation. It has been identified that the mixture of water–ethanol–chloroform as a single-phase solvent is effective for size separation of CPs, where water–ethanol helps to solubilize the CPs but chloroform decreases their solubility. A step-by-step separation was followed using different centrifugation speeds from 4000 to 16,000 rpm, which found that the smallest particle with the highest fluorescence does not precipitate even at 16,000 rpm. All the other size particles show very weak fluorescence and there is very little blue shift in fluorescence with decreasing particle size.

Bhunia et al. (2013) used a variety of carbohydrates and different nucleation-growth conditions to obtain FCNs with tunable emission and with high quantum yield. Figure 3.1 shows the four types of FCN

FCN	Molecular weight/particle size	Elemental composition (C:H:N:O)	Emission$_{max}$ (excitation)	Molar extinction coefficient at excitation	Fluorescence quantum yield
FCN$_{blue}$	400–2200 Da	65:6:8:21	440 nm (370 nm)	2×10^3	6–30%
FCN$_{green}$	2500–14000 Da, 2–4 nm	75:10:5:10	500 nm (400 nm)	5×10^4	14%
FCN$_{yellow}$	1–4 nm	50:15:2:33	560 nm (425 nm)	4×10^3	12%
FCN$_{red}$	~4–10 nm	70:5:1:24	600 nm (385 nm)	7×10^5	7%

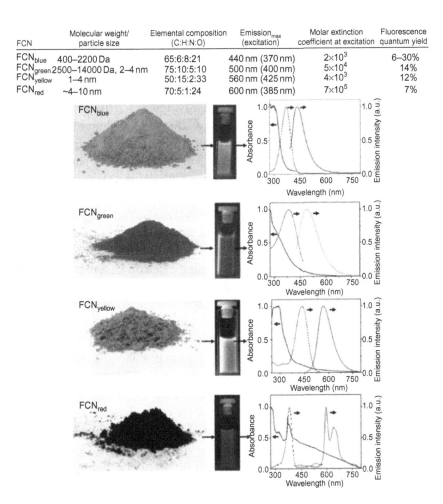

Figure 3.1 Digital image of gram scale solid FCN samples, digital images of their solutions under respective excitations, and their absorption (−), excitation (…), and emission (color lines) spectra. Emission spectra have been measured by exciting at 370 nm for FCN$_{blue}$, by exciting at 400 nm for FCN$_{green}$, by exciting at 425 nm for FCN$_{yellow}$, and by exciting at 385 nm for FCN$_{red}$. All excitation spectra are recorded in respective emission maxima. Also, different properties of these FCNs are tabulated (Bhunia et al., 2013). Copyright 2015 by the Nature Publishing Group.

with different emission colors. Although all samples show excitation-dependent emission spectra, each sample has most intense emissions at certain excitation. Sizes and different physical properties of different FCN are tabulated within Figure 3.1. The TEM study of CPs obtained from candle soot by Ray et al. (2009) shows that synthesized CP have broad size distributions from 20–350 nm, but FCNs have small and narrow particle size distributions from 2 to 6 nm, as shown in Figure 3.2(a) and (b).

For the measurement of TEM, diluted FCN solution was dropped onto copper grids to prepare specimens for transmission electron microscopic (TEM) observation with a field-emission gun operating at 200 kV. Both the CP and FCN have good graphitization, and the interlayer spacing between graphitic sheets is $d_{(002)} = 0.33$ nm as obtained from HRTEM, as shown in the inset of Figure 3.2(b), which is very close to that of the ideal graphite. A similar type of carbon nanoparticle (1.5–2.5 nm) is reported after an aqueous route with the help of a silica sphere as the carrier (Liu et al., 2009). This type of size-dependent fluorescence QY is observed for CP produced via laser ablation technique (Cao et al., 2012) and thermal decomposition (Batalov et al., 2009; Zyubin et al., 2009), and particles are obtained from candle soot (Liu et al., 2007). Ray et al. (2009) observed that the CP and FCN obtained from burning candle soot has a very strong tendency for aggregation during TEM grid preparation or scanning electron microscopy (SEM) slide preparation, as shown in Figure 3.2(a). The aggregation is so high that it is difficult to find a significant amount of isolated small FCNs under TEM either from synthesized CP solution or from FCN solution. A similar type of aggregation was observed by Iijima et al. (1999), who found that small CPs aggregate into ∼80-nm nanohorn structures. Ray et al. (2009) found that the multiple drop samples show larger particles than the single drop. It was estimated that the particle size of CNPs was 12–15 nm for the single drop sample (Figure 3.2(c)), but the size increased when the sample was prepared from three drops (Figure 3.2 (d)). Figure 3.2(e) shows a typical AFM image of FCN structures. Figure 3.3(a) shows the Raman spectra of the FCN and soot, respectively, obtained from Ray et al. (2009). Micro Raman studies were performed using an ISA Lab Raman system equipped with a 514.5-nm laser with a $100 \times$ objective, giving a spot size of ∼1 μm with a spectral resolution better than 2 cm^{-1}. The spectrum of FCNs shows a high photoluminescence background compared with soot.

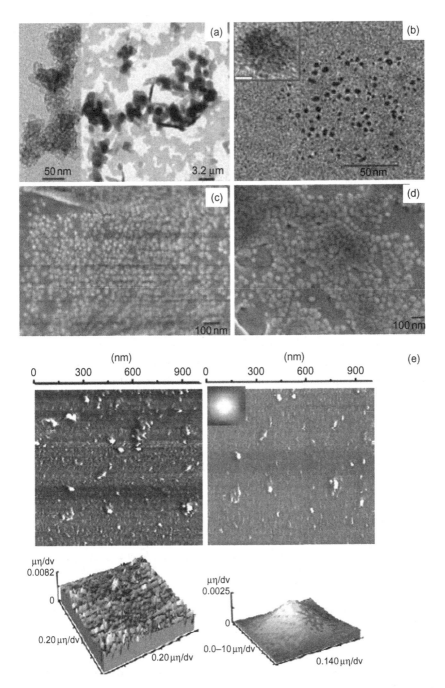

Figure 3.2 (a) TEM images of synthesized CP showing broad size/shape distribution as well as extensive particle agglomeration. (b) TEM images of small FCN with high-resolution image of one particle in the inset. (c) SEM image of single drop of FCN solution onto Si substrate. (d) SEM image of FCN solution on Si substrate but after three successive drops. (e) AFM image of FCN taken at different positions and magnifications. From Ray et al., 2009. Copyright 2015 American Chemical Society.

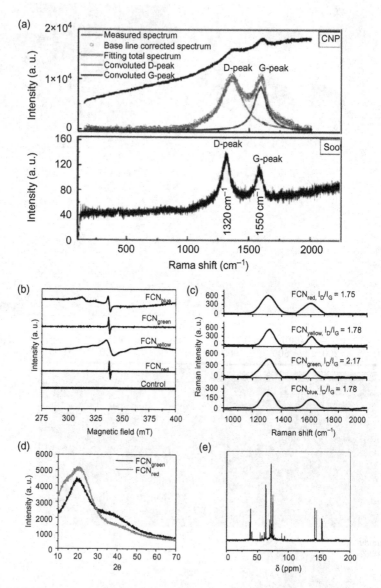

Figure 3.3 (a) Raman spectra of the FCN and raw powder of soot (Ray et al. 2009). (b) EPR spectra of FCN measured at 25°C. (c) Raman spectra of FCNs measured at 785 nm. (d) XRD of FCN$_{green}$ and FCN$_{red}$. (e) ^{13}C NMR spectrum of FCN$_{green}$. From Bhunia et al., 2013. Copyright 2015 Nature publishing group.

The two signature peaks for carbon, i.e., D band and G band, are clearly seen for FCN and soot, where the D band corresponds to the disordered structure of the crystalline sp^2 cluster and the G band corresponds to the in-plane stretching vibration mode E$_{2g}$ of single-crystal graphite. The intensity ratio (I_D/I_G), which is often used to correlate the structural purity to graphite, also indicates that FCN is composed of mainly nanocrystalline graphite (Xia et al., 2004). The size of the nanocrystalline graphite obtained from the relation deduced by Ferrari et al. (2000) was calculated as 2.2 nm. Figure 3.3(b) shows the EPR spectra of different FCNs obtained from carbonization of carbohydrate measured at 25°C in comparison with the control carbon nanoparticles sample having poor fluorescence; however, Figure 3.3(c) shows Raman spectra of different fluorescent FCN measured at 785 nm laser excitation showing prominent D and G bands along with their intensity ratio.

Figure 3.3(d) shows the XRD of FCN_{green} and FCN_{red} obtained by Bhunia et al. (2013) from oxidation of carbohydrates, showing peak at $2\theta \sim 20°$, corresponding to amorphous CPs, and the hump at $2\theta \sim 40°$, corresponding to (100) plane of particles. Figure 3.3(c) shows the ^{13}C NMR spectrum of FCN_{green}, indicating the presence of both sp^3 (corresponding δ values of 8–80 ppm) and sp^2 (corresponding δ values of 90–180 ppm) carbon atoms.

3.3.2 Chemical and Bonding Properties

Composition of FCN has been characterized by different methods such as elemental analysis, X-ray photoelectron microscopy (XPS), Fourier transform infrared spectroscopy (FTIR), and proton NMR spectroscopy. Burning candle soot contains mainly elemental carbon and oxygen, having 96 atomic% and 4 atomic%, respectively, whereas synthesized FCN from candle soot shows C, O, and N of 59 atomic%, 37 atomic%, and 4 atomic%, respectively, as estimated from XPS compositional analysis. Comparative XPS data in Figure 3.4(a)–(f) show that FCN is mainly composed of graphitic carbon (sp^2) and oxygen/nitrogen-bonded carbon, whereas starting soot is mainly composed of diamond-like carbon (sp^3) with oxygen-bonded carbon. FCN synthesized from carbonization of carbohydrates (Bhunia et al., 2013) has carbon as a major component (50–75%), along with other minor components such as hydrogen, nitrogen, oxygen, and phosphorus (Baker and Hammer, 1997; Zheng et al., 1996; Mansour et al., 1993; Robinson, 1974; Paulmier et al., 2007; Klages et al., 2003; Lukaszewicz, 1997; Sato et al., 2004; Choi et al., 2005; Yudasaka et al., 1994, Ghosh et al., 2008; Wei et al., 1999). These compositions vary depending on the nature of precursor carbohydrates and presence of surface-adsorbed molecules, as shown in Figure 3.4 and tabulated in Figure 3.1. In some cases, XPS analysis confirms the

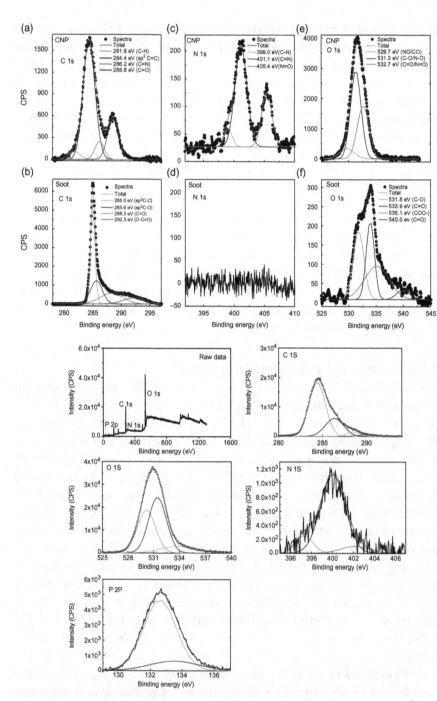

Figure 3.4 (Top) XPS spectra of (a) C 1s, (c) N 1s, and (e) O 1s of FCN [C:O:N = 59:37:4]; and (b) C 1s, (d) N 1s, and (f) O 1s of raw candle soot [C:O = 96:4] (Ray et al., 2009). (Bottom) Raw data and deconvoluted C 1s, O 1s, N 1s, and P 2p XPS of FCN_red (Bhunia et al., 2013). Copyright 2015 American Chemical Society and Nature publishing group.

presence of trace of nitrogen/phosphorus in some of the FCNs, which are not detectable by conventional CHN analysis. In the XPS study, C-1s, N-1s, and O-1s spectra of CNP obtained from candle soot deconvolution into different peaks was executed by curve fitting using Gaussian functions. The peaks were assigned to $C-N/C=N$ and $C-H$ bonds (Mansour et al., 1993; Robinson, 1974; Paulmier et al., 2007), $-C=O$ carbonyl groups (Klages et al., 2003; Lukaszewicz, 1997), $C-C$ (Sato et al., 2004), $C-O/C-H$, carbonyl groups (Sato et al., 2004), CO_2, and/or $C-C=O$ bonds (Choi et al., 2005). In case of N 1s, the peaks are assigned as $C-N$ and $C=N$ bond (Yudasaka et al., 1994; Ghosh et al., 2008; Wei et al., 1999), $N-O$, and/or $N=O$ bonds (Sato et al., 2004). The O1s peak is assigned as $C-O/N-O$, $C=O/N=O$, $C-O$, $-C=O$, and COOH (Wei et al., 1999). The composition variation obtained from XPS measurements is well-matched with bonding structures obtained from the FTIR spectrum (Xia et al., 2004; Mutsukura et al., 1999). In FTIR spectra, the peaks at ~ 1100 and $1265\,cm^{-1}$ are $C-O-C$ and $C=C$ bonds, respectively. A few small peaks are observed between 1400 and $1650\,cm^{-1}$ and at $1735\,cm^{-1}$, which are ascribed to $C-H$, conjugated $C-N$ and $C=N$, and $N-O$ bonds of stretching modes, respectively. The broad band observed at $1600-1700\,cm^{-1}$ can be assigned to the $C-OH$ and $C=O$ bonds. A small band at $1835\,cm^{-1}$ is observed and is associated with $C=N$ (sp^2 $C-N$) stretching vibration. The band observed at approximately $2090\,cm^{-1}$ is ascribed to the $C-H$ bond. The bands 2335 and $2360\,cm^{-1}$ can be attributed to the nitrile $(-CN)$ group. In addition, a broad IR band $\sim 3100-3650\,cm^{-1}$ appeared due to $C-OH$ and COOH bonds. It is well-known that nitric acid oxidation produces OH and CO_2H groups on the carbon nanoparticle surfaces that made them hydrophilic and negatively charged particles (Boehm, 1994). In addition, this oxidation can also induce, to a small extent, nitration into graphitic carbon (Salame et al., 2001; Kamegawa et al., 2002). The experimental data suggested that the refluxing step with nitric acid has made two-fold chemical modifications to the soot. First, it induces partial oxidation of carbons and introduces functional groups such as OH, CO_2H, and NO_2. Second, it induces doping of nitrogen and oxygen into the CP.

Introduction of functional groups induces water solubility and surface charge to the FCN. In addition, it helps to break the large aggregated soot particle into small CPs. This oxidation step can also be considered as a chemical route of incorporating nitrogen and oxygen into the CP as observed from the chemical composition analysis. Both proton NMR and FTIR studies show the signature of surface functional groups like OH, CH, COOH, NH, and NH_2. The magnetic measurement study does not show any significant magnetic property. Lifetime decay of all the FCNs fits with two to three component lifetimes in the time scale $0.1-10$ ns, suggesting multiple radiative species are present in each FCNs.

3.3.3 Optical-Luminescence Properties

The fluorescence (photoluminescence) spectra were measured using fluorescence spectrophotometer at different excitation energy. Ray et al. (2009) showed the fluorescence spectra of FCNs obtained from the burning of candle soot at different excitation energy ranging from $325-600$ nm, as shown in Figure 3.5. Ray et al. (2009) found that the highest fluorescence intensity was achieved at the excitation wavelength of 450 nm, which shows maximum emission at 520 nm and is green emission (as shown in Figure 3.5). As a comparison, UV–VIS absorption spectra of aqueous solution of synthesized CP and the smallest FCNs of $2-6$ nm are shown in Figure 3.5, where CP shows clear absorption near 325 nm. Figure 3.5 (bottom) shows the different FCN photoluminescence spectra obtained by Bhunia et al. (2013) at different excitation wavelengths. Ray et al. also tested the quality and yield of CP as a function of reflux time and found that 12 h is the optimum time (Ray et al., 2009). If the reflux time is less, the yield is low; for longer reflux time, the yield does not increase appreciably (Ray et al., 2009). Ray et al. (2009) also observed that the intensity of fluorescence of various phosphate-buffered solutions of FCNs does not change appreciably on solution pH from 7 to 9. To extend the application potential of FCN as a fluorescent label, Bhunia et al. (2013) have investigated the fluorescence stability under different conditions, as shown in Figure 3.6 (top). It shows the fluorescence of FCN solutions under different pH conditions, indicating that fluorescence is stable in a wider range of pH and under continuous excitations.

In this case, the films of FCNs are prepared by depositing respective aqueous solutions on the glass slides and then imaged under fluorescence mode with blue/green excitation before and after 60-min excitation with 365-nm UV lamp. Figure 3.6 (bottom) shows lifetime study of FCN and shows that the value is in the range of the nanosecond time scale. In Figure 3.6 (bottom), black/green lines represent

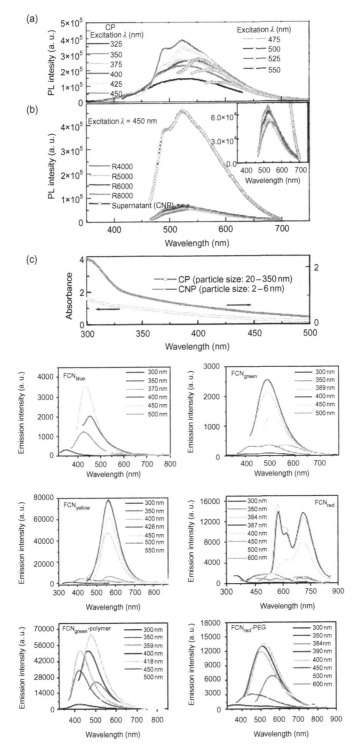

Figure 3.5 (Top) Fluorescence spectra of (a) synthesized CP, (b) different sizes of CP obtained at different centrifugations (intense spectrum for smallest size [2–6 nm]), (c) UV–visible absorption spectra of aqueous solution of synthesized CP and smallest FCNs (2–6 nm) (Ray et al., 2009). (Bottom) Emission spectra of different FCN (Bhunia et al, 2013). Copyright 2015 American Chemical Society and Nature Publishing Group.

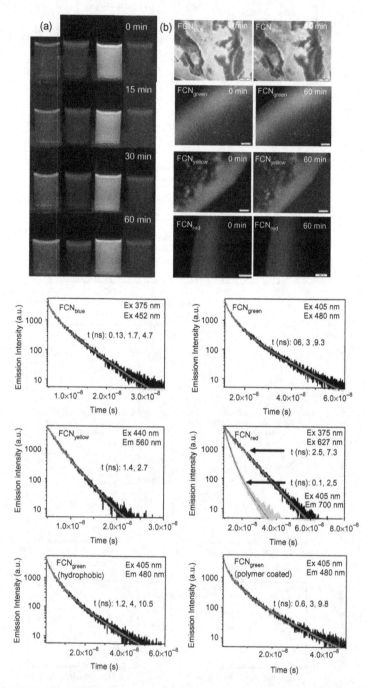

Figure 3.6 (Top) Photostability studies of (a) different FCN solutions and (b) different FCN films (Bhunia et al., 2013). (Bottom) Fluorescence lifetime decay spectra of solutions of different FCN (Bhunia et al., 2013). Copyright 2015 Nature Publishing Group.

experimental data and red/blue lines correspond to fitted data. Experimental excitation/emission wavelength and resultant lifetime values are shown inside each graph.

3.4 APPLICATION OF CARBON NANOPARTICLES IN BIOIMAGING PROCESS

Water-soluble FCN is considered as ideal cell imaging probe with minimum cytotoxicity (Li, H. et al., 2010; Li, Q. et al., 2010; Guo et al., 2012; Zyubin et al., 2009). However, functionalization is an important step for cellular and subcellular targeting (Lm et al., 2009). Interestingly, Ray et al. (2009) found that FCNs enter into cells without any further functionalization, and using the fluorescence property of FCN it is possible to track them, as shown in Figure 3.7 (top). Typically, FCN solution is mixed with cell culture media along with cells and incubated for 30 min, and washed cells are then imaged under bright field, UV, and blue excitations. Cells become bright blue–green under UV excitations and yellow under blue excitation, but they are colorless in the control sample where no FCNs were used. This suggests that FCNs enter into cells and labeled cells can be imaged using a conventional fluorescence microscope. In this case, cell solution was mixed with FCN solution and incubated for 30 min. Washed cells were imaged under bright field, UV, and blue excitations. The bottom row images correspond to the control experiment in which no FCN was used. Cells become bright blue–green under UV excitations and yellow under blue excitation, but they were colorless in the control sample. A light blue color of the control sample under UV excitation is due to well-known auto-fluorescence of cells. Functionalized FCN-based cell labeling is shown in Figure 3.7 (middle and bottom). Figure 3.7 (middle) shows the bioimaging performance of oleyl functionalized FCN$_{green}$; bright field and fluorescence images of oleyl functionalized FCN$_{green}$-labeled COS-7 cells are shown in the left panel and right panel. Oleyl functional group has been covalently linked with carbon nanoparticle, and it offers increased cell membrane targeting and cell uptake (Medintz et al., 2005). It is observed that oleyl functionalization of FCN has significantly increased the cellular interaction/uptake; however, without any oleyl functionalization, the labeling was insignificant. Bhunia et al. (2013) have transformed different types of FCN into different functional FCN using conventional coating and conjugation chemistry, and have shown their potential as biological labels.

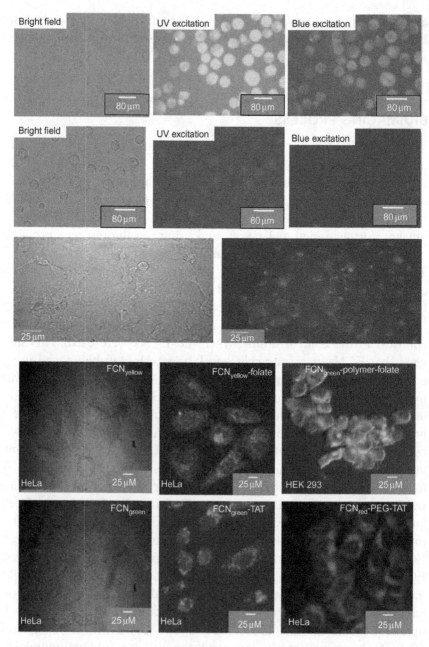

Figure 3.7 (Top) FCN-based labeling of EAC cells (Ray et al., 2009). (Middle) Bright field (left panel) and fluorescence (right panel) images of oleyl functionalized FCN$_{green}$-labeled COS-7 cells. (Bottom) Fluorescence imaging of cells labeled with functional FCN. FCNs were incubated with cells for 3–6 h and labeled cells were imaged under fluorescence microscope. Cells were imaged under confocal fluorescence microscope or Apotome microscope (Bhunia et al., 2013). Copyright 2015 American Chemical Society and Nature publishing group.

Figure 3.7 (bottom) shows some other examples of cell images labeled with different functional FCN. In this work, hydrophobic FCNs have been transformed into water-soluble FCN via the well-established amphiphilic polymer coating approach (Yang et al., 2012; Bourlinos et al., 2008a,b).

In these processes, lipophilic polymers are coated with the fatty amine-capped hydrophobic FCN and transformed into water soluble, primary amine-terminated FCN. Next, they are functionalized with different affinity molecules via conventional conjugation chemistry. Hydrodynamic sizes of these functional FCNs are relatively large (typically 5–15 nm) compared with synthesized FCN because of the coating layer of polymer. In contrast, hydrophilic FCNs can be directly transformed into functional FCNs using the primary amine/carboxylate group present on their surface. Such amine/acid functional groups are either from monomer precursors or from capping ligands. This approach has also been used to prepare TAT peptide and folate-functionalized FCNs via conjugation chemistry (Bourlinos et al., 2008a,b; Wang, F. et al., 2010; Wang, X. et al., 2010). It was found that this approach produces functional FCNs with small hydrodynamic diameters (<10 nm) and is more appropriate for bioimaging applications. All these functional FCNs retained fluorescence for different biomedical applications. To prove that the functional FCNs are useful as imaging probes, Bhunia et al. (2013) have investigated their performance as fluorescent cell labels, as shown in Figure 3.7 (bottom). TAT peptide or folate-functionalized FCNs are mixed with cell culture medium with cells attached on the cultured plates; after incubation for several hours, washed cells are imaged under a fluorescence microscope. Cells get labeled within 1–2 h, and labeled cells can be imaged with a conventional fluorescence microscope. Control experiments show that FCNs have very low nonspecific binding to cells due to small hydrodynamic diameter and low surface charge (Pan et al., 2010). In addition, TAT functionalization increases the cell labeling/uptake and folate functionalization offers selective labeling of cells that have folate receptors (Pan et al., 2010).

3.5 CYTOTOXICITY OF FCN

Cytotoxicity of FCN has been studied extensively and some of the results are shown in Figure 3.8. Cytotoxicity study has been performed

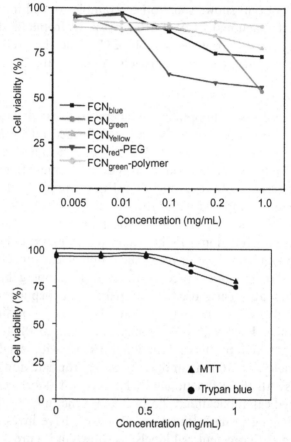

Figure 3.8 (Top) MTT-based cytotoxicity assay for different FCN, showing that they are nontoxic at substantially high doses as compared with the usual required concentration (<0.2 mg/mL) for cell labeling. Some toxicity for FCN_green and FCN_red-PEG are observed at high doses, possibly due to primary amine groups present on their surface. (Bottom) Cytotoxicity of FCN for HepG2 cell studied by MTT and Trypan blue assays. All data are the average of three experiments (Bhunia et al., 2013). Copyright 2015 American Chemical Society and Nature Publishing Group.

using methylthiazolyldiphenyl-tetrazolium bromide (MTT) assay and Trypan blue assay. Typically, cells are exposed to FCN for 24 h in the concentration range of 0.1−1 mg/mL, which is ∼100- to 1000-times higher than required for imaging application. Next, MTT solution was added to each well 4 h before the end of the incubation. The medium was discarded and the produced formazan was dissolved with DMSO. The plates were read with absorbance at 550 nm. The optical density is directly correlated with cell quantity, and cell viability was calculated by assuming 100% viability in the control set without any CNP. In case of Trypan blue assay, 0.4% of Trypan blue solution was used instead of MTT, and after 5 min stained cells were counted to determine the cell viability. Results show that cell survival rates are very high in the tested concentrations. However, at higher concentrations some percentage of cell death is observed, which is mainly due to related surface chemistry. This result concludes that FCN can be used in high concentrations for imaging or other biomedical applications.

3.6 DISCUSSION

We have tried to correlate the origin of tunable emission and high fluorescence quantum yield with particle size and chemical composition of FCN. Our results show that there is a correlation between particle size and emission color. As the size of FCN increases the absorption edge, excitation maxima and emission maxima are redshifted. The maximum size limit in obtaining fluorescent CP is typically $\sim 5-10$ nm, and thus synthetic methods that produce carbon nanoparticles >10 nm are either nonfluorescent or weakly fluorescent. However, size is not the only factor in dictating the emission. Presence of other elements, such as oxygen, nitrogen, and phosphorus, and their relative ratios in the carbon matrix dictate the quantum yield (see Table in Figure 3.1). In addition, there exists a correlation between existence of defect sites and emission. Presence of defect sites is evident from ESR signal corresponding to free electrons and high ratio of Raman D-to-G band. The high ratio of sp^3 carbon and presence of oxygen, nitrogen, and phosphorus would enhance the defect sites (Cao et al., 2012). Based on these observations, a tentative explanation of tunable emission can be explained as follows: with the increasing particle size, the extent of the sp^2 domain in the sp^3 matrix increases. This increase in sp^2 domain increases the conjugated double bonds. The role of oxygen, nitrogen, and phosphorus is to control sp^2 domain by controlling the extent of sp^3 carbon matrix and/or by controlling the defect sites (Cao et al., 2012). Thus, key features of the present synthetic method are restriction of the CP size $<5-10$ nm and effective incorporation of other elements in small size particles that induce defect sites. The presented FCN-based probes are far superior in property as compared with other nanoparticle-based probes reported previously. Gold nanoparticle-based probes, which offer dark field imaging, have limited applications in cellular imaging because of the high scattering background from cells. Semi-conductor nanocrystal (or quantum dot)-based fluorescent probes are limited by toxic cadmium present in their composition. Fluorescent nanoprobes based on doped semi-conductor nanocrystals, fluorescent gold clusters, and fluorescent silicon nanoparticles are under development and have yet to prove whether their performance is similar to quantum dots. In contrast, FCN-based probes are composed of nontoxic carbon materials with tunable fluorescence and hydrodynamic diameter <10 nm. We and others have convincingly demonstrated that the fluorescence of FCN is size-dependent, fluorescence is stable during chemical modification, and various functional FCNs with tunable emission can be prepared similar to quantum dots. It has also been demonstrated that functional FCN can be used as fluorescent biological labels to image/detect molecules inside cells. These labeling applications indicate that FCNs can be a powerful nontoxic alternative to semi-conductor nanocrystals. In addition, smaller size of FCNs should be more powerful for subcellular targeting and renal clearance, which are often limited for semi-conductor nanocrystals. Synthetic and functionalization approaches presented here produce high-quality FCNs in milligram to gram scales, along with options for scaling up. Thus, developed approaches would provide large-scale availability of high-quality FCNs and could test them for various biomedical applications.

In the soot-based approach, yield of soluble CP depends of the oxidation property of the reagent (e.g., nitric acid), and the fluorescent quantum yield of CP seems to depend on the efficiency of nitrogen and oxygen incorporation. Smaller particle size and dominant graphitic structure of the raw soot made this oxidation step easier. However, the efficiency of converting soot into soluble CP is still low (yield ~20%), as observed from large amounts of insoluble soot. This suggests that this type of chemical oxidation is not efficient enough for complete conversion into water-soluble particles. Longer time refluxing with nitric acid increases the yield of soluble particles to some extent, but does not increase the yield of fluorescent CP. This suggests that further oxidation might have other adverse effects, such as further oxidation of CP that reduces the conjugated double bond structure in the CP.

Incorporation of nitrogen and oxygen defects via nitric acid oxidation might have a role in producing the fluorescent center in CP. Such defect structures in the fluorescence property of diamond are well-established (Baker et al., 2010; Li, H. et al., 2010; Li, Q. et al., 2010). Because soot has some percentage of diamond-like carbon, as identified from XPS measurements, it is possible that during the oxidation step, nitrogen and oxygen defects are formed into diamond structure. However, their presence in FCN is too low to determine with our presented analytical method and further study is needed to confirm this possibility. An alternative explanation of fluorescence may be that chemical oxidation and the doping step introduces more conjugated double bond system into the CP, thereby introducing the fluorescence property. Nevertheless, the advantage of this type of chemical process of making fluorescent CP is that it is simple and requires less adverse conditions as compared with ion beam radiation method (Baker et al., 2010; Li, H. et al., 2010; Li, Q. et al., 2010). However, because synthesized CPs have heterogeneous size distribution, small particles are more fluorescent than larger ones. Thus, successful isolation of small particles is essential for the enhancement of fluorescence quantum yield.

3.7 CONCLUSION AND PERSPECTIVES OF CARBON NANOPARTICLES

FCNs 1–10 nm in diameter were obtained by different researchers using different oxidation and carbonization processes using soot particles and carbohydrates. For the oxidation process, the surface oxidation and subsequent nitrogen and oxygen doping afforded light-emitting properties of CP. It is to be noted that the light emitted by these CPs depends on the wavelength of light used for excitation. Different sizes of CPs were isolated and emission quantum yield was size-dependent, i.e., the smaller the size, the better the photoluminescence efficiency. The fluorescence property of these particles is useful for cell imaging application. These FCNs enter into cells without any further functionalization, and fluorescence properties of the particle can be used to track their position in cells using a conventional fluorescence microscope. The discovery of FCNs will no doubt lead to more research in this field; these particles have potential in biomedical applications where cadmium-based quantum dots show toxic effects. In case of CPs obtained from the carbonization process of carbohydrates, high FCNs with tunable visible emission from blue to red are synthesized in gram scale and transformed into various biological labels. Small size, stable emission, and low toxicity of these nanoprobes give them an advantage over widely used semi-conductor nanocrystals. Future work should focus on utilizing the full potential of these nanomaterials, particularly in biomedical science.

3.8 PRESENT CHALLENGES AND FUTURE RESEARCH IN CARBON NANOPARTICLES

Although significant progress has been made in this area, further development is needed with a particular focus on several areas. First, the current status of understanding the origin of fluorescence is poor and multiple mechanisms have been proposed. This lack of clear understanding limits further advancement of synthetic methods. For example, if the doping of heteroatoms is the main reason for high-fluorescence quantum yield, then synthetic conditions should be focused on. This understanding would also help to develop high-quality red/NIR-emitting FCNs. Second, functionalization of FCNs is not yet advanced and only few functional FCNs have been reported. Knowledge advanced in the area of graphene functionalization should be utilized and scientists with an organic chemistry background should be involved. Third, more detailed studies should be conducted on FCN-induced in vitro and in vivo cytotoxicity. This would provide more understanding of the long-term effects of FCNs on the environment and on human consumption. Fourth, more in vivo studies should be conducted to exploit the advantages of FCN. This study should include targeting different organs and the brain. With these advancements, it is expected that this research area would create new excitements in the next 5–10 years.

REFERENCES

Baker, M.A., Hammer, P., 1997. A study of the chemical bonding and microstructure of ion beam-deposited CN_x films including an XPS C 1s peak simulation. Surf. Interface Anal. 25, 629–642.

Baker, S.N., et al., 2010. Luminescent carbon nanodots: emergent nanolights. Angew. Chem. Int. Ed. 49, 6726–6744.

Bao, L., et al., 2011. Electrochemical tuning of luminescent carbon nanodots: from preparation to luminescence mechanism. Adv. Mater. 23, 5801–5806.

Batalov, A., et al., 2009. Low temperature studies of the excited-state structure of negatively charged nitrogen-vacancy color centers in diamond. J. Phys. Rev. Lett. 102 (195506), 1–4.

Bhunia, S.K., et al., 2013. Carbon nanoparticle-based fluorescent bioimaging probes. Sci. Rep. 3 (1473), 1–7.

Boehm, H.P., 1994. Some aspects of the surface chemistry of carbon blacks and other carbons. Carbon 32, 759–769.

Bourlinos, A.B., et al., 2008a. Photoluminescent carbogenic dots. Chem. Mater. 20, 4539–4541.

Bourlinos, A.B., et al., 2008b. Surface functionalized carbogenic quantum dots. Small 4, 455–458.

Cahalan, M.D., et al., 2002. Real-time imaging of lymphocytes in vivo. J. Nat. Rev. Immunol. 2, 872–880.

Cao, L., et al., 2007. Carbon dots for multiphoton bioimaging. J. Am. Chem. Soc. 129, 11318–11319.

Cao, L., et al., 2013. Photoluminescence properties of grapheme versus other carbon nanomaterials. Acc. Chem. Res. 46 (1), 171–180.

Chandra, S., et al., 2012. Tuning of photoluminescence on different surface functionalized carbon quantum dots. RSC Adv. 2, 3602–3606.

Choi, H.C., et al., 2005. Distribution and structure of N atoms in multiwalled carbon nanotubes using variable-energy X-ray photoelectron spectroscopy. J. Phys. Chem. B 109, 4333–4340.

Diederich, F., Thilgen, C., 1996. Covalent fullerene chemistry. Science 271, 317–323.

Dreyer, D.R., et al., 2010. The chemistry of graphene oxide. Chem. Soc. Rev. 39, 228–240.

Fang, Y., et al., 2012. Easy synthesis and imaging applications of cross-linked green fluorescent hollow carbon nanoparticles. ACS Nano 6, 400–409.

Fu, C.C., et al., 2007. Characterization and application of single fluorescent nanodiamonds as cellular biomarkers. Proc. Natl. Acad. Sci. USA 104, 727–732.

Ghosh, P., et al., 2008. Bamboo-shaped aligned CN_x nanotubes synthesized using single feedstock at different temperatures and study of their field electron emission. J. Phys. D: Appl. Phys. 41 (155405), 1–7.

Glinka, Y.D., et al., 1999. Multiphoton-excited luminescence from diamond nanoparticles. J. Phys. Chem. B 103, 4251–4263.

Gruber, A., et al., 1997. Scanning confocal optical microscopy and magnetic resonance on single defect centers. Science 276, 2012–2014.

Guo, X., et al., 2012. Facile access to versatile fluorescent carbon dots toward light-emitting diodes. Chem. Commun. 48, 2692–2694.

Hens, S.C., et al., 2012. Photoluminescent nanostructures from graphite oxidation. J. Phys. Chem. C 116, 20015–20022.

Huang, L.C., et al., 2004. Adsorption and immobilization of cytochrome c on nanodiamonds. Langmuir 20, 5879–5884.

Iijima, S., et al., 1999. Nano-aggregates of single-walled graphitic carbon nano-horns. Chem. Phys. Lett. 309, 165–170.

Jana, N.R., et al., 2004. Size- and shape-controlled magnetic (Cr, Mn, Fe, Co, Ni) oxide nanocrystals via a simple and general approach. Chem. Mater. 16, 3931–3935.

Kamat, P.V., 2008. CdSe quantum dot sensitized solar cells. Semiconductor nanocrystals as light harvesters. J. Phys. Chem. C 112, 18737–18753.

Kamegawa, K., et al., 2002. Oxidative degradation of carbon blacks with nitric acid: II. Formation of water-soluble polynuclear aromatic compounds. Carbon 40, 1447–1455.

Klages, K.U., et al., 2003. Deuterium bombardment of carbon and carbon layers on titanium. J. Nucl. Mater. 313, 56–61.

Kong, X., et al., 2005. Polylysine-coated diamond nanocrystals for MALDI-TOF mass analysis of DNA oligonucleotides. Anal. Chem. 77, 4273–4277.

Kong, X.L., et al., 2005. High-affinity capture of proteins by diamond nanoparticles for mass spectrometric analysis. Anal. Chem. 77, 259–265.

Krysmann, M.J., et al., 2012. Formation mechanism of carbogenic nanoparticles with dual photoluminescence emission. J. Am. Chem. Soc. 134, 747–750.

Li, H., et al., 2010. Water-soluble fluorescent carbon quantum dots and photocatalyst design. Angew. Chem. Int. Ed. 49, 4430–4434.

Li, Q., et al., 2010. Photoluminescent carbon dots as biocompatible nanoprobes for targeting cancer cells in vitro. J. Phys. Chem. C 114, 12062–12068.

Li, H., et al., 2011. One-step ultrasonic synthesis of water-soluble carbon nanoparticles with excellent photoluminescent properties. Carbon 49, 605–609.

Lim, T.-S., et al., 2009. Fluorescence enhancement and lifetime modification of single nanodiamonds near a nanocrystalline silver surface. Phys. Chem. Chem. Phys. 11, 1508–1514.

Liu, H., Ye, T., et al., 2007. Fluorescent carbon nanoparticles derived from candle soot. Angew. Chem. Int. Ed. 46, 6473–6475.

Liu, R., et al., 2009. An aqueous route to multicolor photoluminescent carbon dots using silica spheres as carriers. Angew. Chem. Int. Ed. 48, 4598–4601.

Lukaszewicz, J.P., 1997. X-ray photoelectron spectroscopy studies of sodium modified carbon films suitable for use in humidity sensors. J. Mater. Sci. 32, 6063–6068.

Mansour, A., et al., 1993. Photoelectron-spectroscopy study of amorphous a-CN_x:H. Phys. Rev. B 47, 10201–10209.

Medintz, I.L., et al., 2005. Quantum dot bioconjugates for imaging, labelling and sensing. Nat. Mater. 4, 435–446.

Mochalin, V.N., Gogotsi, Y., 2009. Wet chemistry route to hydrophobic blue fluorescent nanodiamond. J. Am. Chem. Soc. 131, 4594–4595.

Mutsukura, N., et al., 1999. Infrared absorption spectroscopy measurements of amorphous CN_x films prepared in CH_4/N_2 r.f. discharge. Thin Solid Films 349, 115–119.

Neugart, F., et al., 2007. Dynamics of diamond nanoparticles in solution and cells. Nano Lett. 7, 3588–3591.

Pan, D., et al., 2010. Observation of pH-, solvent-, spin-, and excitation-dependent blue photoluminescence from carbon nanoparticles. Chem. Commun. 46, 3681–3683.

Paulmier, T., et al., 2007. Deposition of nano-crystalline graphite films by cathodic plasma electrolysis. Thin Solid Films 515, 2926–2934.

Peng, H., Travas-Sejdic, J., 2009. Simple aqueous solution route to luminescent carbogenic dots from carbohydrates. Chem. Mater. 21, 5563–5565.

Peng, J., et al., 2012. Graphene quantum dots derived from carbon fibers. Nano Lett. 12, 844–849.

Ray, S.C., et al., 2009. Fluorescent carbon nanoparticles: synthesis, characterization, and bioimaging application. J. Phys. Chem. C 113, 18546–18551.

Robinson, J.W., 1974. Handbook of Spectroscopy, vol. 1. CRC Press, Cleveland, OH.

Salame, I.I., et al., 2001. Surface chemistry of activated carbons: combining the results of temperature-programmed desorption, Boehm, and potentiometric titrations. Colloid Interface Sci. 240, 252–258.

Sato, T., et al., 2004. Preparation of carbon nitride film by cryogenic laser processing. Appl. Phys. A 79, 1477–1479.

Selvi, B.R., et al., 2008. Intrinsically fluorescent carbon nanospheres as a nuclear targeting vector: delivery of membrane-impermeable molecule to modulate gene expression in vivo. Nano Lett. 8, 3182–3188.

Sun, X., Li, Y., 2004. Colloidal carbon spheres and their core/shell structures with noble metal nanoparticles. Angew. Chem. Int. Ed. 43, 597–601.

Sun, Y.P., et al., 2006. Quantum-sized carbon dots for bright and colorful photoluminescence. J. Am. Chem. Soc. 128, 7756–7757.

Tasis, D., et al., 2006. Chemistry of carbon nanotubes. Chem. Rev. 106, 1105–1136.

Ushizawa, K., et al., 2002. Covalent immobilization of DNA on diamond and its verification by diffuse reflectance infrared spectroscopy. Chem. Phys. Lett. 351, 105–108.

Wang, F., et al., 2010. One-step synthesis of highly luminescent carbon dots in non-coordinating solvents. Chem. Mater. 22, 4528–4530.

Wang, X., et al., 2010. Bandgap-like strong fluorescence in functionalized carbon nanoparticles. Angew. Chem. Int. Ed. 49, 5310–5314.

Wang, X., et al., 2011. Microwave assisted one-step green synthesis of cell-permeable multi-colour photoluminescent carbon dots without surface passivation reagents. J. Mater. Chem. 21, 2445–2450.

Wee, T.-L., et al., 2007. Two-photon excited fluorescence of nitrogen-vacancy centers in proton-irradiated type Ib diamond. J. Phys. Chem. A 111, 9379–9386.

Wei, J., et al., 1999. TEM, XPS and FTIR characterization of sputtered carbon nitride films. Surf. Interface Anal. 1999 (28), 208–211.

Xia, Y., et al., 2004. Synthesis of ordered mesoporous carbon and nitrogen-doped carbon materials with graphitic pore walls via a simple chemical vapor deposition method. Adv. Mater. 16, 1553–1558.

Yang, Z.-C., et al., 2011. Intrinsically fluorescent nitrogen-containing carbon nanoparticles synthesized by a hydrothermal process. Carbon 49, 5207–5212.

Yang, Y., et al., 2012. One-step synthesis of amino-functionalized fluorescent carbon nanoparticles by hydrothermal carbonization of chitosan. Chem. Commun. 48, 380–382.

Yu, S.J., et al., 2005. Bright fluorescent nanodiamonds: no photobleaching and low cytotoxicity. J. Am. Chem. Soc. 127, 17604–17605.

Yu, B.Y., et al., 2012. Carbon quantum dots embedded with mesoporous hematite nanospheres as efficient visible light-active photocatalysts. J. Mater. Chem. 22, 8345–8353.

Yudasaka, M., et al., 1994. Graphite thin film formation by chemical vapor deposition. Appl. Phys. Lett. 64, 842–844.

Zhang, B., et al., 2010. A novel one-step approach to synthesize fluorescent carbon nanoparticles. Eur. J. Inorg. Chem. 26, 4411–4414.

Zhang, J., et al., 2010. Controlled synthesis of green and blue luminescent carbon nanoparticle with high yields by the carbonization of sucrose. New J. Chem. 34, 591–593.

Zhao, Q.-L., et al., 2008. Facile preparation of low cytotoxicity fluorescent carbon nanocrystal by electrooxidation of graphite. Chem. Commun. 41, 5116–5118.

Zheng, W.T., et al., 1996. Reactive magnetron sputter deposited CN_x: effects of N_2 pressure and growth temperature on film composition, bonding, and microstructure. J. Vac. Sci. Technol. A 14, 2696–2701.

Zhou, J., et al., 2007. An electrochemical avenue to blue luminescent nanocrystals from multi-walled carbon nanotubes (MWCNTs). J. Am. Chem. Soc. 129, 744−745.

Zhu, A., et al., 2012. Carbon-dot-based dual-emission nanohybrid produces a ratiometric fluorescent sensor for in vivo imaging of cellular copper ions. Angew. Chem. Int. Ed. 51, 7185−7189.

Zhuo, S., et al., 2012. Upconversion and downconversion fluorescent graphene quantum dots: ultrasonic preparation and photocatalysis. ACS Nano 6, 1059−1064.

Zyubin, A.S., et al., 2009. Quantum chemical modeling of photoabsorption properties of two- and three-nitrogen vacancy point defects in diamond. J. Phys. Chem. C 113, 10432−10440.

Printed in the United States
By Bookmasters